GCSE IN A WEEK

MATHS

Fiona Mapp

Revision Planner

Page	Day	Time (mins)	Title	Exam Board	Date	Time	Completed
Day 1							
4	1	10 mins	Prime factors, HCF and LCM	AEO			
6	1	10 mins	Fractions and Surds	AEO			
8	1	10 mins	Percentages	AEO			
10	1	15 mins	Repeated percentage change	AEO			
12	1	15 mins	Reverse percentage problems	AEO			
14	1	10 mins	Ratio and Proportion	AEO			
Day 2							
16	2	15 mins	Indices	AEO			
18	2	10 mins	Standard index form	AEO			
20	2	20 mins	Proportionality	AEO			
22	2	15 mins	Upper and Lower bounds	AEO			
24	2	15 mins	Formulae and Expressions	AEO			
26	2	15 mins	Brackets and Factorisation	AEO			
28	2	15 mins	Equations 1	AEO			
Day 3							
30	3	10 mins	Equations 2	AEO			
32	3	15 mins	Solving quadratic equations 1	AEO			
34	3	15 mins	Solving quadratic equations 2	AEO			
36	3	15 mins	Simultaneous linear equations	AEO			
38	3	15 mins	Solving linear and quadratic equations simultaneously	AEO			
Day 4							
40	4	15 mins	Algebraic fractions	AEO			
42	4	10 mins	Sequences	AEO			
44	4	10 mins	Inequalities	AEO			
46	4	15 mins	Straight-line graphs	AEO			
48	4	15 mins	Curved graphs	AEO			
50	4	20 mins	Interpreting graphs	AEO			
52	4	15 mins	Functions and Transformations	AEO			

Prime factors, HCF and LCM

Any number can be written as a product of prime factors. This means the number is written using only prime numbers multiplied together.

The prime numbers up to 20 are:

2, 3, 5, 7, 11, 13, 17, 19

The diagram below shows the prime factors of 60.

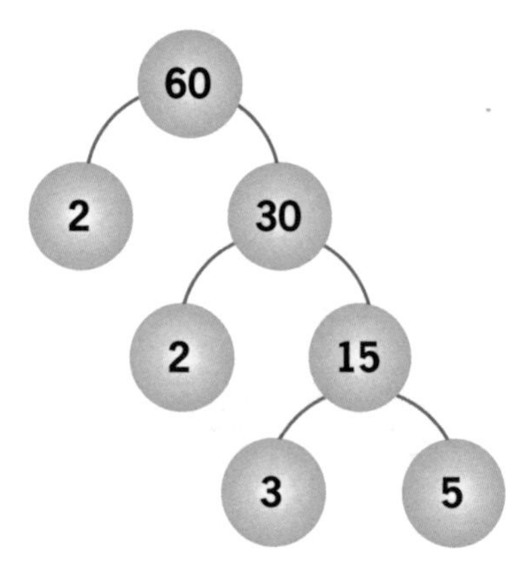

- Divide 60 by its first prime factor 2
- Divide 30 by its first prime factor 2
- Divide 15 by its first prime factor 3
- We can now stop because the second number 5 is prime.

As a product of its prime factors 60 may be written as:

$60 = 2 \times 2 \times 3 \times 5$

or

$60 = 2^2 \times 3 \times 5$

Highest Common Factor (HCF)

The largest factor that two numbers have in common is called the HCF.

Example

Find the HCF of 60 and 96.

- Write the numbers as the product of their prime factors.

$60 = \boxed{2} \times \boxed{2} \times \boxed{3} \times 5$

$96 = \boxed{2} \times \boxed{2} \times 2 \times 2 \times 2 \times \boxed{3}$

- Ring the factors that are in common.
- These give the HCF $= 2 \times 2 \times 3$

$= 12$

Lowest Common Multiple (LCM)

This is the lowest number that is a multiple of two numbers.

Example

Find the LCM of 60 and 96.

- Write the numbers as products of their prime factors.

$$60 = 2 \times 2 \times 3 \times 5$$

$$96 = 2 \times 2 \times 2 \times 2 \times 2 \times 3$$

- 60 and 96 have a common prime factor of $2 \times 2 \times 3$, so it is only counted once.
- The LCM of 60 and 96 is

$2 \times 2 \times 2 \times 2 \times 2 \times 3 \times 5$

$= 480$

1. Write these numbers as a product of prime factors:
   a. 50
   b. 360
   c. 16
2. Decide whether these statements are true or false.
   a. The HCF of 20 and 40 is 4.
   b. The LCM of 6 and 8 is 24.
   c. The HCF of 84 and 360 is 12.
   d. The LCM of 24 and 60 is 180.
3. Find the HCF and LCM of 36 and 48.

EXAM QUESTION

1. Express 120 as the product of powers of its prime factors.
2. Find the Lowest Common Multiple of 120 and 42.

Fractions and Surds

A fraction is part of a whole one. The top number is the numerator, and the bottom number is the denominator.

FOUR RULES OF FRACTIONS

Addition

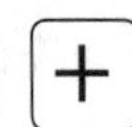

$\frac{5}{9} + \frac{1}{7}$

- You need to change the fractions so that they have the same denominator.

$= \frac{35}{63} + \frac{9}{63}$

$= \frac{44}{63}$

- Remember to only add the numerators and not the denominators.

Multiplication

- Before starting, write out whole or mixed numbers as top-heavy fractions.

$\frac{2}{7} \times \frac{4}{5}$

$\frac{2 \times 4}{7 \times 5}$

Multiply the numerators together

Multiply the denominators together

$= \frac{8}{35}$

Division

- Before starting, write whole or mixed numbers as top-heavy fractions.

$2\frac{1}{3} \div 1\frac{2}{7}$

$= \frac{7}{3} \div \frac{9}{7}$

- Convert to top heavy fractions.

$= \frac{7}{3} \times \frac{7}{9}$

- Take reciprocal of second fraction and multiply.

$= \frac{49}{27}$

$= 1\frac{22}{27}$

- Rewrite as a mixed number.

Subtraction

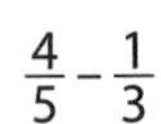

$\frac{4}{5} - \frac{1}{3}$

$= \frac{12}{15} - \frac{5}{15}$

$= \frac{7}{15}$

- Change the fractions into their equivalents with a denominator of 15.

Manipulating surds

Irrational numbers involving square roots are also called surds.

When working with surds there are several rules to learn:

1. $\sqrt{a} \times \sqrt{b} = \sqrt{ab}$

   e.g. $\sqrt{3} \times \sqrt{5} = \sqrt{15}$

2. $(\sqrt{b})^2 = \sqrt{b} \times \sqrt{b} = b$

   e.g. $(\sqrt{5})^2 = \sqrt{5} \times \sqrt{5} = 5$

3. $\frac{\sqrt{a}}{\sqrt{b}} = \sqrt{\frac{a}{b}}$

   e.g. $\frac{\sqrt{10}}{\sqrt{2}} = \sqrt{\frac{10}{2}} = \sqrt{5}$

4. $(a + \sqrt{b})^2$

   $= (a + \sqrt{b})(a + \sqrt{b})$

   $= a^2 + 2a\sqrt{b} + (\sqrt{b})^2$

   $= a^2 + 2a\sqrt{b} + b$

5. $(a + \sqrt{b})(a - \sqrt{b})$

   $= a^2 - a\sqrt{b} + a\sqrt{b} - (\sqrt{b})^2$

   $= a^2 - b$

   e.g. $(2 + \sqrt{3})(2 - \sqrt{3}) = 4 - 3$

   $= 1$

Exam Question

1. Work out $\frac{(5 + \sqrt{5})(2 - 2\sqrt{5})}{\sqrt{45}}$.

   Give your answer in its simplest form.

2. Rationalise the denominator.

   $\frac{1}{\sqrt{2}}$

Rationalising the denominator

Sometimes a surd can appear on the bottom of the fraction. It is usual to rewrite the surd so that it appears as the numerator. This is called **rationalising** the denominator.

Example

Rationalise $\frac{4}{\sqrt{7}}$

$= \frac{4}{\sqrt{7}} \times \frac{\sqrt{7}}{\sqrt{7}}$

Multiply the top and bottom of the fraction by the surd function. In this case $\sqrt{7}$.

$= \frac{4\sqrt{7}}{(\sqrt{7})^2}$

$= \frac{4\sqrt{7}}{7}$

Progress Check

1. Work out the following:

   a. $\frac{2}{3} + \frac{1}{5}$

   b. $\frac{6}{7} - \frac{1}{3}$

   c. $\frac{2}{9} \times \frac{5}{7}$

   d. $\frac{3}{11} \div \frac{22}{27}$

2. Express the following in the form $a\sqrt{b}$.

   a. $\sqrt{24}$

   b. $\sqrt{200}$

   c. $\sqrt{48} + \sqrt{12}$

Percentages

Percentages are fractions with a denominator of 100.

This is the percentage sign

%

Percentage of a quantity

OF means multiply

Example 1

Find 30% of 80 kg.

$\frac{30}{100} \times 80 = 24\text{ kg}$

On the calculator, key in:

30% = 0.3 – this is known as the **multiplier**

For the **non-calculator** paper:

- find 10% by dividing by 10

  10% of 80 kg

  $= 80 \div 10$

  $= 8\text{ kg}$

- then multiply by 3 to get 30%

  3×8

  $= 24\text{ kg}$

Example 2

A CD player costs £65. In a sale it is reduced by 15%. Work out the cost of the CD player in the sale.

Method 1

15% of £65

$= \frac{15}{100} \times 65$

= £9.75

Price of CD player in the sale

= £65 – £9.75

= £55.25

Method 2
(using a multiplier)

$1 - 0.15 = 0.85$ (0.85 is the multiplier)

0.85×65

= £55.25

One quantity as a percentage of another

To make the answer a percentage –
multiply by 100%

Example 3

Matthew got 46 out of 75 in a Science test.
What percentage did he get?

$\frac{46}{75} \times 100\%$	Make a fraction.
$= 61.3\%$	Multiply by 100%.

On the calculator, key in:

Progress Check

1. Without using a calculator work out the following:

   a. 20% of 50 kg b. 30% of £2000

   c. 5% of £60 d. 35% of 720 g

2. Without using a calculator, change each fraction into a percentage:

   a. $\frac{16}{50}$ b. $\frac{46}{200}$

   c. $\frac{15}{20}$ d. $\frac{21}{25}$

3. A PlayStation costs £225. In a sale it is reduced by 20%. Work out the sale price of the PlayStation.

4. The cost of a rail ticket increases by 15%. If the original cost is £25, work out the new cost of the ticket.

5. Thomas got 27 out of 35 in a test. What percentage did Thomas get?

Exam Question

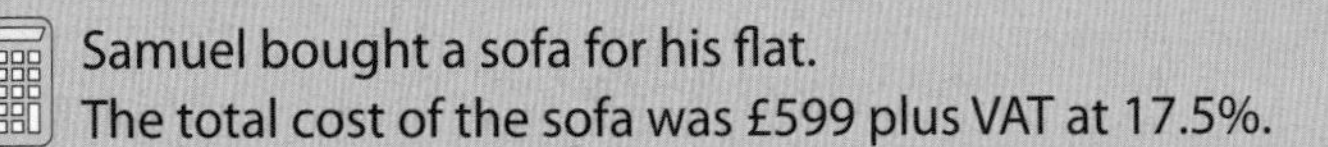

Samuel bought a sofa for his flat.
The total cost of the sofa was £599 plus VAT at 17.5%.

1. Work out the cost of the sofa.

Samuel invited 40 people to a party. Only 32 of the people invited came to the party.

2. Express 32 as a percentage of 40.

Repeated percentage change

Compound interest is where the bank pays interest on the interest earned as well as on the original money.

Percentage change

$$\text{Percentage change} = \frac{\text{change}}{\text{original}} \times 100\%$$

Example

Tammy bought a flat for £185 000. Three years later she sold it for £242 000. What is her percentage profit?

Profit = £242 000 – £185 000

= £57 000

$$\text{Percentage profit} = \frac{57\,000}{185\,000} \times 100\%$$

= 30.8% (3sf)

Example

Jackie bought a car for £12 500 and sold it two years later for £7250. Work out her percentage loss.

Loss = £12 500 – £7250

= £5250

$$\text{Percentage Loss} = \frac{5250}{12\,500} \times 100\%$$

= 42%

Repeated percentage change

A car was bought for £12 500. Each year it depreciated in value by 15%. What was the car worth after three years?

You must remember not to do: 3 × 15% = 45% reduction over 3 years!

Method 1

- Find 100 – 15 = 85% of the value of the car first.

  Year 1 $\frac{85}{100} \times £12\,500 = £10\,625$
- Then work out the value year by year. (£10 625 depreciates in value by 15%.)

  Year 2 $\frac{85}{100} \times £10\,625 = £9031.25$

  (£9301.25 depreciates in value by 15%.)

  Year 3 $\frac{85}{100} \times £9031.25 = £7676.56$

Method 2

- A quick way to work this out is by using a multiplier.
- Finding 85% of the value of the car is the same as multiplying by 0.85.

  Year 1: 0.85 × £12 500

  Year 2: 0.85 × £10 625

  Year 3: 0.85 × £9031.25
- This is the same as working out $(0.85)^3 \times £12\,500 = £7676.56$

Compound interest

Charlotte has £3200 in her savings account and compound interest is paid at 3.2% p.a. How much will she have in her account after four years?

100 + 3.2 = 103.2%

= 1.032   This is the multiplier.

Year 1: 1.032 × £3200 = £3302.40

Year 2: 1.032 × £3302.40 = £3408.08

Year 3: 1.032 × £3408.08 = £3517.14

Year 4: 1.032 × £3517.14 = £3629.68

Total = £3629.68

A quicker way is to multiply £3200 by $(1.032)^4$

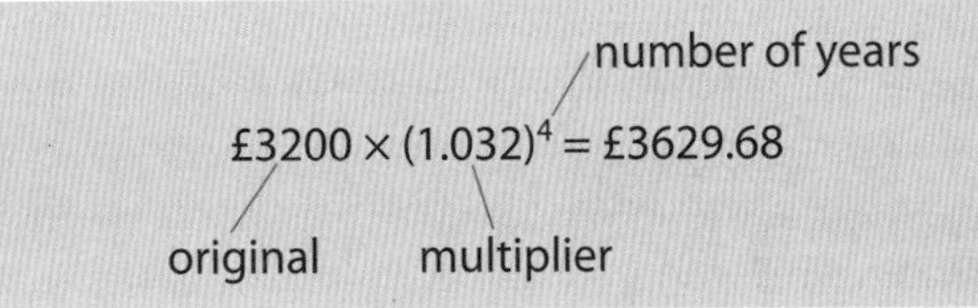

PROGRESS CHECK

1. Imran bought a car for £8500. Two years later he sold it for £4105. Work out his percentage loss.
2. John bought a bike for £135. Six months later he sold it for £95. Work out his percentage loss.
3. Reece has £5200 in the bank. If compound interest is paid at 2% p.a., how much will he have in his account after 3 years?
4. Mr Singh bought a flat for £85 000 in 1999. The flat rose in value by 12% in 2000 and 28% in 2001. How much was the flat worth at the end of 2001?

EXAM QUESTION

1. Sophie buys a new car, costing £13 500. The value of the car depreciates by 10% each year.
   a. Sophie says 'after 10 years the car will have no value'. Sophie is wrong. Explain why.
   b. Sophie wants to work out the value of the car after 2 years. By what single decimal number should Sophie multiply the value of the car when new?
   c. Work out the value of Sophie's car after 2 years.

Reverse percentage problems

These are when the original quantity is calculated. They are quite tricky so think carefully!

The price of a television is reduced by 20% in the sales. It now costs £350. What was the original price?

- The sale price is 100% – 20% = 80% of the pre-sale price (x)
- 80% = 0.8 This is the multiplier.
- $0.8 \times x = £350$

$$x = \frac{£350}{0.8}$$

Original price = £437.50

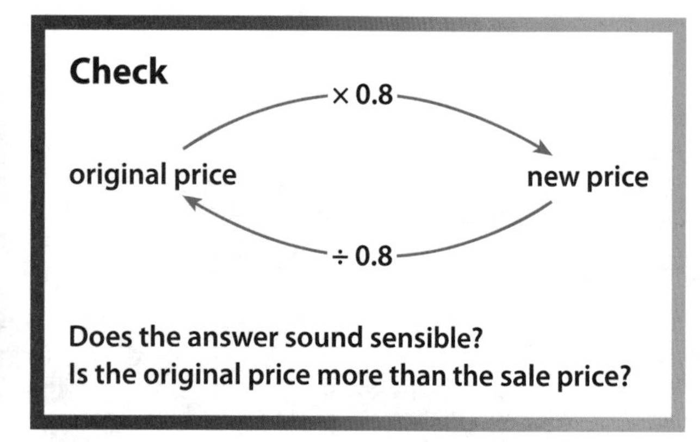

A telephone bill costs £169.20 including VAT at 17.5%. What is the cost of the bill without the VAT?

- The telephone bill of £169.20 represents 100% + 17.5% = 117.5% of the original bill (x).
- 117.5% = 1.175 This is the multiplier.
- $1.175 \times x = £169.20$

$$x = \frac{£169.20}{1.175}$$

Original bill = £144

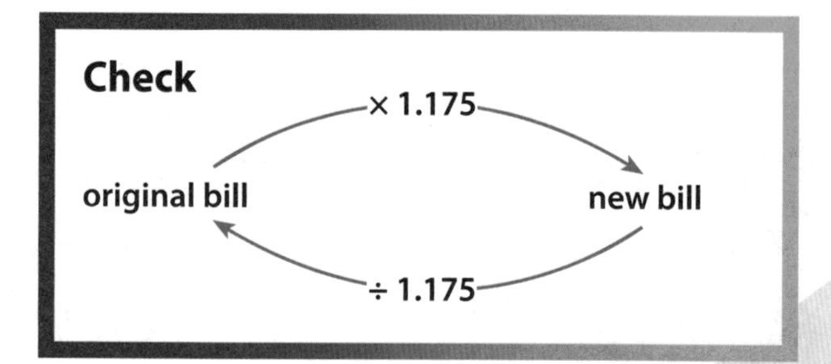

The price of a washing machine is reduced by 15% in the sales. It now costs £323. What was the original price?

- The sale price is 100% – 15% = 85% of the pre-sale price (x).
- 85% = 0.85 — This is the multiplier.
- $0.85 \times x = £323$

$$x = \frac{£323}{0.85}$$

Original price = £380

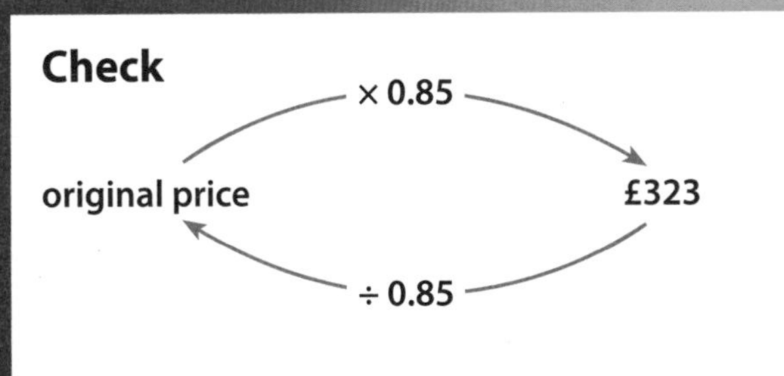

PROGRESS CHECK

1. Each item listed below includes VAT at 17.5%. Work out the original price of the item.

   a. a pair of shoes: £62

   b. a coat: £125

   c. a suit: £245

   d. a TV: £525

2. Joseph says that the original price of a CD player, which now costs £60 after a 15% reduction, was £70.59. Is Joseph correct?

EXAM QUESTION

1. In a sale, normal prices are reduced by 12%. The sale price of a television is £220.

   Work out the normal price of the television.

Ratio and Proportion

A ratio is used to compare two or more related quantities.

Sharing a quantity in a given ratio

To share an amount into proportional parts, add up the individual parts and then divide the amount by this number to find one part.

Example
£155 is divided in the ratio of 2 : 3 between Daisy and Tom. How much does each receive?

$2 + 3 = 5$ parts — Add up the total parts.

5 parts = £155

1 part = £155 ÷ 5 — Work out what one part is worth.

= £31

So Daisy gets 2 × £31 = £62
and Tom gets 3 × £31 = £93.

Check: £62 + £93

= £155 ✔

Best buys

Use unit amounts to help you decide which is the better value for money.

Example
The same brand of breakfast cereal is sold in two different sized packets. Which packet represents better value for money?

- Find the cost per gram for both boxes of cereal.

  125 g costs £1.65 so 165 ÷ 125
  = 1.32 p per gram

  500 g costs £3.15 so 315 ÷ 500
  = 0.63 p per gram

- Since the larger box costs less per gram, it represents better value for money.

Increasing and decreasing in a given ratio

A patio took 4 builders 6 days to build.

At the same rate how long would it take 6 builders?

Time for 4 builders $= 6$ days

Time for 1 builder $= 6 \times 4 = 24$ days

It takes 1 builder four times as long to build the patio.

Time for 6 builders $= 24 \div 6 = 4$ days

Example

A photograph of length 12 cm is to be enlarged in the ratio 4 : 5.

What is the length of the enlarged photograph?

$12 \div 4 = 3$ cm	Divide 12 by 4 to get 1 part.
$3 \times 5 = 15$ cm	Multiply this by 5 to get the length of the enlarged photograph.

1. Divide £160 in the ratio 1 : 2 : 5
2. The cost of four ringbinders is £6.72. Work out the cost of 21 ringbinders.
3. Write 45 : 60 as a ratio in its simplest form.
4. 80 g of plain flour is used to make an apple cake for 4 people. How much plain flour is needed to make an apple cake for 14 people?
5. A wall took 6 builders 4 days to build. At the same rate how long would it take 8 builders?

1. Jenny, Amy and Robert share £108 in the ratio 3 : 4 : 2. Calculate the amount of money that Amy receives.
2. Here is a recipe for making 15 chocolate chip cookies.

Chocolate chip cookies
makes 15 cookies
150 g of flour
90 g of sugar
75 g of margarine
60 g of chocolate chips
3 eggs

Work out the amounts needed to make 25 chocolate chip cookies.

....................g of flour ....g of sugar

....................g of margarine ....g of chocolate chips

....................eggs

Indices

An index is sometimes called a power.

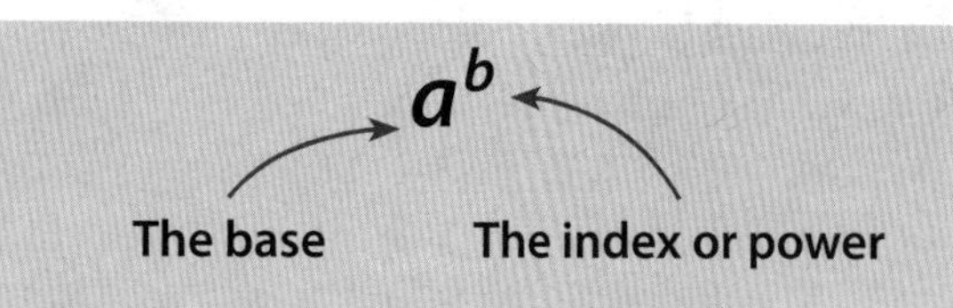

The **base** has to be the **same** when the laws of indices are applied. The laws of indices can be used for numbers or in algebra.

Laws of indices

$$a^n \times a^m = a^{n+m}$$

$$a^n \div a^m = a^{n-m}$$

$$(a^n)^m = a^{n \times m}$$

$$a^0 = 1$$

$$a^1 = a$$

$$a^{-n} = \frac{1}{a^n}$$

$$a^{\frac{1}{m}} = \sqrt[m]{a}$$

$$a^{\frac{n}{m}} = (\sqrt[m]{a})^n$$

Examples with numbers

1. Simplify the following, leaving your answer in index notation.

   a. $5^2 \times 5^3 = 5^{2+3} = 5^5$

   b. $8^{-5} \times 8^{12} = 8^{-5+12} = 8^7$

   c. $(2^3)^4 = 2^{3\times4} = 2^{12}$

2. Evaluate: Evaluate means to work out.

   a. $4^2 = 4 \times 4 = 16$

   b. $5^0 = 1$

   c. $3^{-2} = \frac{1}{3^2} = \frac{1}{9}$

   d. $36^{\frac{1}{2}} = \sqrt{36} = 6$

   e. $8^{\frac{2}{3}} = (\sqrt[3]{8})^2 = 2^2 = 4$

3. Simplify the following, leaving your answer in index form.

   a. $7^2 \times 7^5 = 7^7$

   b. $6^9 \div 6^2 = 6^7$

   c. $3^{-7} \times 3^6 = 3^{-1} = \frac{1}{3}$

   d. $7^9 \div 7^{-10} = 7^{19}$

4. Evaluate:

   a. $3^3 = 3 \times 3 \times 3 = 27$

   b. $7^0 = 1$

   c. $63^{\frac{1}{3}} = \sqrt[3]{64} = 4$

   d. $81^{\frac{1}{2}} = \sqrt{81} = 9$

   e. $5^{-2} = \frac{1}{5^2} = \frac{1}{25}$

Examples with algebra

1. Simplify the following:

   a. $a^4 \times a^{-6} = a^{4-6} = a^{-2} = \frac{1}{a^2}$

   b. $5y^2 \times 3y^6 = 15y^8$

   The numbers are multiplied — The indices are added

   c. $(4x^3)^2 = 16x^6$

   Remember to square the 4 as well.

   d. $(2x)^{-3} = \frac{1}{(2x)^3} = \frac{1}{8x^3}$

2. Simplify:

   a. $\frac{15b^4 \times 3b^7}{5b^2} = \frac{45b^{11}}{5b^2} = 9b^9$

   b. $\frac{16a^2b^4}{4ab^3} = 4ab$

3. Simplify:

   a. $7a^2 \times 3a^2b = 21a^4b$

   b. $\frac{14a^2b^4}{7ab} = 20ab^3$

   c. $\frac{9x^2y \times 2xy^3}{6xy} = \frac{18x^3y^4}{6xy}$

   $= 3x^2y^3$

1. Simplify the following, leaving your answers in index form.

   a. $6^3 \times 6^5$

   b. $12^{10} \div 12^{-3}$

   c. $7^{10} \div 7^{-14}$

   d. $(5^2)^3$

   e. $64^{\frac{5}{6}}$

2. Simplify the following:

   a. $2b^4 \times 3b^6$

   b. $8b^{-12} \div 4b^4$

   c. $(3b^4)^2$

   d. $\frac{9b^6 \times 2b^5}{3b^{-3}}$

   e. $(5x^2y^3)^{-2}$

EXAM QUESTION

1. Evaluate

   a. 5^0

   b. 7^{-2}

   c. $64^{\frac{1}{3}}$

   d. $27^{\frac{-2}{3}}$

Standard index form

Standard index form is used to write very large or very small numbers in a simpler way.

When written in standard form a number will be written as

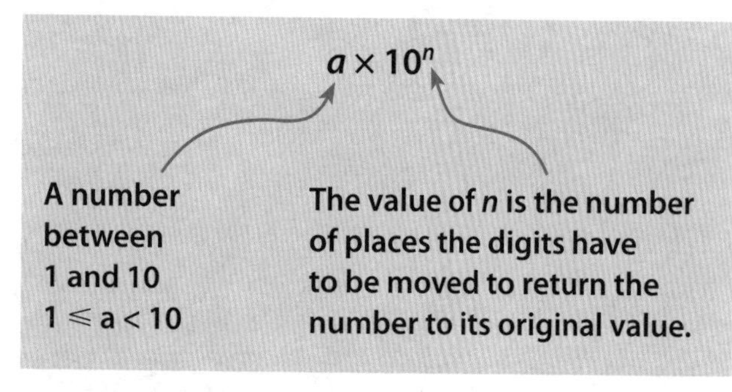

Example

Write 2 730 000 in standard form.

2.73 is the number between 1 and 10 ($1 \leqslant 2.73 < 10$).

Count how many spaces the digits have to move to restore the original number.

The digits have moved 6 places to the left because it has been multiplied by 10^6

2.73

2 7 3 0 0 0 0

So $2\,730\,000 = 2.73 \times 10^6$

Write 0.000046 in standard form.

4.6×10^{-5}

- Put the decimal point between the 4 and 6, so the number lies between 1 and 10.

Move the digits five places to the right to restore the original number.

The value of n is negative.

- **If the number is greater than 1, n is positive.**
- **If the number is less than 1, n is negative.**

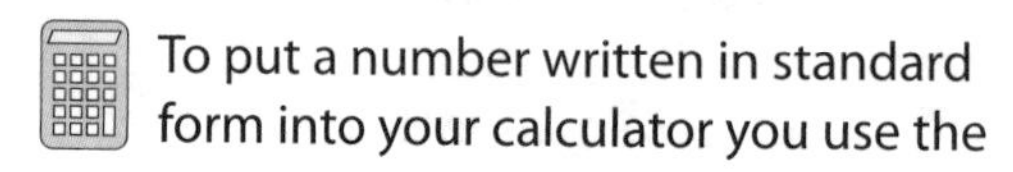

To put a number written in standard form into your calculator you use the

EXP or EE key.

$(2 \times 10^3) \times (6 \times 10^7) = 1.2 \times 10^{11}$

This would be keyed in as

2 EXP 3 × 6 EXP 7 =

Example

Work out the following using a calculator. Check that you get the answer given.

a. $(6.7 \times 10^7)^3 = 3.0 \times 10^{23}$

b. $(4 \times 10^9) \div (3 \times 10^4)^2 = 4.\dot{4}$

c. $\dfrac{(5.2 \times 10^6) \times (3 \times 10^7)}{(4.2 \times 10^5)^2} = 884.4$

Example
On a non-calculator paper you can use indices to help work it out.

a. $(2 \times 10^3) \times (6 \times 10^7)$
$= 2 \times 6 \times 10^3 \times 10^7$
$= 12 \times 10^{3+7}$
$= 12 \times 10^{10}$
$= 1.2 \times 10 \times 10^{10}$
$= 1.2 \times 10^{11}$

b. $(6 \times 10^4) \div (3 \times 10^{-2})$
$= (6 \div 3) \times (10^4 \div 10^{-2})$
$= 2 \times 10^{4-(-2)}$
$= 2 \times 10^6$

c. $(3 \times 10^4)^2$
$= (3 \times 10^4) \times (3 \times 10^4)$
$= 3 \times 3 \times 10^4 \times 10^4$
$= 9 \times 10^8$

PROGRESS CHECK

1. Write in standard form:
   a. 64 000
   b. 271 000
   c. 0.00046
   d. 0.000000074
2. Without a calculator, work out the following. Leave in standard form.
   a. $(3 \times 10^4) \times (4 \times 10^6)$
   b. $(6 \times 10^{-5}) \div (3 \times 10^{-4})$
   c. $(5 \times 10^6) \times (7 \times 10^9)$
3.  Work these out on a calculator:
   a. $(4.6 \times 10^{12}) \div (3.2 \times 10^{-6})$
   b. $(7.4 \times 10^9)^2 + (4.1 \times 10^{11})$

EXAM QUESTION

1. a. Write 40 000 000 in standard form.
   b. Write 6×10^{-5} as an ordinary number.
2. Work out the value of $6 \times 10^{-5} \times 40\,000$.
   Give your answer in standard form.

Proportionality

The notation $\propto$ means 'is directly proportional to'. This is often abbreviated to 'is proportional to' or 'varies as'.

Direct proportion

As one variable increases, the other increases and as one variable decreases the other decreases.

Example

If a is proportional to the square of b and $a = 5$ when $b = 4$, find the value of k (the constant of proportionality) and the value of a when $b = 8$.

1. Change the sentence into proportionality using the symbol $\propto$.

$a \propto b^2$

2. Replace $\propto$ with "$= k$" to make an equation.

$a = kb^2$

3. Substitute the value given in the question in order to find k. Rearrange the equation $5 = k \times 4^2$

$\frac{5}{16} = k$

4. Replace k with the value just found.

$a = \frac{5}{16}b^2$

If $b = 8$ $\quad a = \frac{5}{16} \times 8^2$

$a = \frac{5}{16} \times 64$

$= 20$

Example

A train is accelerating out of a station at a constant rate.

The distance, d metres, gone from the station varies directly as the square of the time taken, t seconds.

After 2.5 seconds the train has gone 20 m.

a. Work out the formula connecting d and t.

$d \propto t^2$

$d = kt^2$

$20 = k \times 2.5^2$

$k = \frac{20}{2.5^2}$

$= 3.2$

$d = 3.2t^2$

b. How long does the train take to go 100 m?

$100 = 3.2t^2$

$t^2 = \frac{100}{3.2}$

$= 31.25$

$t = \sqrt{31.25}$

$t = 5.59$ seconds

Inverse proportion

- As one variable increases the other decreases, and as one variable decreases the other increases.
- If y is inversely proportional to x then we write this as $y \propto \frac{1}{x}$ or else $y = \frac{k}{x}$.

Example

p is inversely proportional to the cube of w. If $w = 2$ when $p = 5$, what is the value of w when $p = 10$?

$p \propto \frac{1}{w^3}$ — Write the information with the proportionality sign.

$p = \frac{k}{w^3}$ — Replace with the constant of proportionality.

$5 = \frac{k}{2^3}$

$5 \times 8 = k$

$k = 40$ — Find the value of k.

$p = \frac{40}{w^3}$ — Rewrite the equation.

$10 = \frac{40}{w^3}$ — find the value of w if $p = 10$.

$w^3 = 4$

$w = \sqrt[3]{4}$

$w = 1.587$

PROGRESS CHECK

1. Match these statements with the correct equation.

a. y is proportional to x	$y = \frac{k}{x}$
b. y is inversely proportional to cube root of x	$y = kx^3$
c. y is inversely proportional to x	$y = kx$
d. y is proportional to x^3	$y = \frac{k}{\sqrt[3]{x}}$

2. a is proportional to $\sqrt{x}$. When $x = 4$, $a = 8$. What is the equation of proportionality and what is the value of x when $a = 64$?

EXAM QUESTION

1. y is inversely proportional to the square of x. When $x = 3$, $y = 10$.

   a. Find an expression for y in terms of x.

   b. Calculate y when $x = 2$

   c. Calculate x when $y = 6$

Upper and Lower bounds

Measurements are never exact. They can only be expressed to a certain degree of accuracy.

When measurements are quoted to a given unit, say nearest metre, there is a highest and lowest value they could be.

Highest value – called the **Upper bound**

Lowest value – called the **Lower bound**

6.2 cm rounded to the nearest millimetre would really lie between

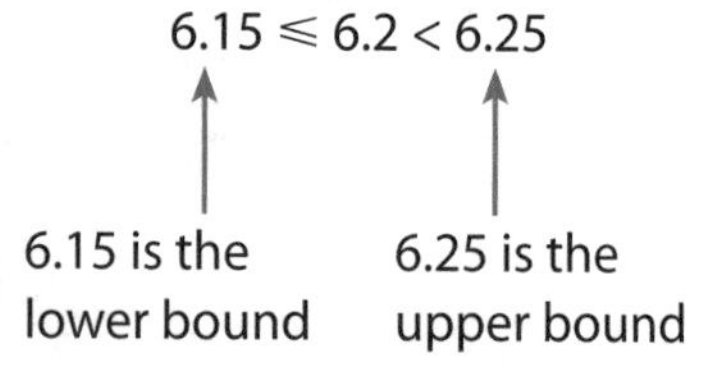

The real value can be as much as half the rounded unit below or above the rounded off value.

Finding maximum and minimum possible values of a calculation

Some exam questions may ask you to calculate the upper or lower bounds of calculations.

TIPS

	Upper bound	**Lower bound**
Addition	Upper bound + Upper bound	Lower bound + Lower bound
Multiplication	Upper bound × Upper bound	Lower bound × Lower bound
Subtraction	(Upper bound of larger quantity) − (Lower bound of smaller quantity)	(Lower bound of larger quantity) − (Upper bound of smaller quantity)
Division	$\frac{\text{Upper bound of quantity 1}}{\text{Lower bound of quantity 2}}$	$\frac{\text{Lower bound of quantity 1}}{\text{Upper bound of quantity 2}}$

Example

1. A square measures 62 mm to the nearest mm. Work out the upper and lower area of the square (in mm^2).

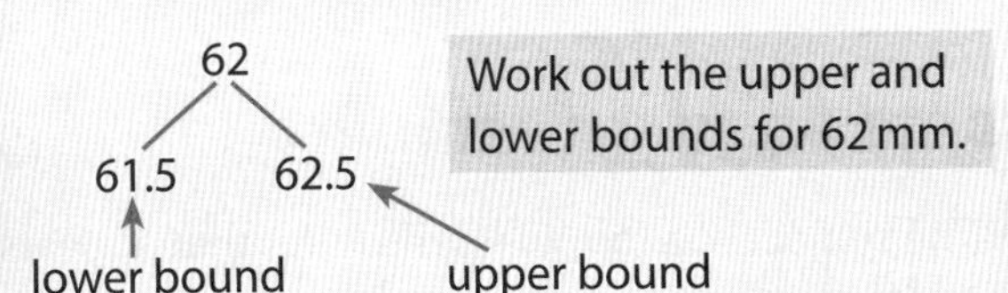

Upper bound $= 62.5 \times 62.5$

$= 3906.25\,\text{mm}^2$

Lower bound $= 61.5 \times 61.5$

$= 3782.25\,\text{mm}^2$

2. Given that $a = 4.2$ (to 1dp) and $b = 6.23$ (to 3sf), find the upper and lower bounds for the following calculations:

a. $b - a$ b. $\dfrac{b}{a}$

4.2 → 4.15, 4.25 6.23 → 6.225, 6.235

a. Upper bound $= 6.235 - 4.15 = 2.085$

Lower bound $= 6.225 - 4.25 = 1.975$

b. Upper bound $= \dfrac{6.235}{4.15} = 1.5024$

Lower bound $= \dfrac{6.225}{4.25} = 1.4647$

PROGRESS CHECK

1. $c = \dfrac{(2.6)^3 \times 12.52}{3.2}$

2.6 and 3.2 are correct to 1 decimal place. 12.52 is correct to 2 decimal places. Which of the following calculations give the lower bound for c and the upper bound for c?

A $\dfrac{(2.65)^3 \times 12.525}{3.15}$

B $\dfrac{(2.55)^3 \times 12.515}{3.15}$

C $\dfrac{(2.65)^3 \times 12.525}{3.25}$

D $\dfrac{(2.65)^3 \times 12.515}{3.15}$

E $\dfrac{(2.55)^3 \times 12.515}{3.25}$

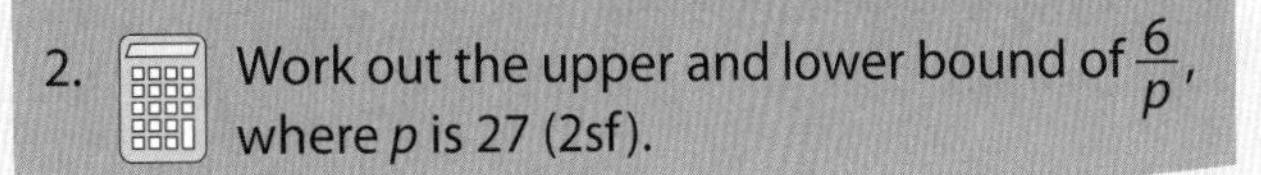

2. Work out the upper and lower bound of $\dfrac{6}{p}$, where p is 27 (2sf).

EXAM QUESTION

1. The value of R is calculated by using this formula:

$$R = \frac{a - b}{b}$$

$a = 7.65$ correct to 2 decimal places

$b = 4.3$ correct to 1 decimal place.

Find the difference between the lower bound of R and the upper bound of R.

Formulae and Expressions

$a + b$ is called an **expression**. $b = a + 6$ is called a **formula**. The value of b depends on the value of a.

- A term is a collection of numbers, letters and brackets, all multiplied together, e.g. $6a$, $2ab$, $3(x–1)$
- Terms are separated by + and – signs. Each term has a + or – sign in front of it.

$3ab \; - \; 4c \; + \; 3b^2 \; + \; 2$

invisible plus sign → before $3ab$; ab term → $3ab$; c term → $4c$; b^2 term → $3b^2$; number term → 2

$3ab$ means $3 \times a \times b$
$3b^2$ means $3 \times b \times b$

Substituting into formulae

Replacing a letter with a number is called **substitution.**

- Write out the expression first and then replace the letters with the values given.
- Work out the value – but take care with the order of operations, i.e. BIDMAS.

Example

1. $a = 3b - 4c$
   Find a if $b = 4$ and $c = -2$.
   $a = (3 \times 4) - (4 \times -2)$
   $= 12 - (-8)$
   $= 20$

   Taking away a negative is the same as adding.

2. $E = \frac{1}{2}mv^2$

   Find E if $m = 6$ and $v = 10$.

   $E = \frac{1}{2} \times 6 \times 10^2$

   $E = 300$

Rearranging formulae

- The subject of a formula is the letter that appears on its own on one side of the formula.

Example

1. Make a the subject of the formula

   $b = (a - 3)^2$.

   $\pm\sqrt{b} = a - 3$ — Deal with the power first, square root both sides.

   $\pm\sqrt{b} + 3 = a$ — Remove any term added or subtracted. Add 3 to both sides.

   $a = \pm\sqrt{b} + 3$ or $a = 3 \pm \sqrt{b}$

2. Make x the subject of the formula

   $p = x^2 + y$

   $p - y = x^2$ — Subtract y from both sides.

   $\pm\sqrt{p - y} = x$ — Square root both sides.

   $x = \pm\sqrt{p - y}$

Example

Make t the subject of the formula $v = u + at$

$v = u + at$	
$v - u = at$	Subtract u from both sides.
$\frac{v - u}{a} = t$	Divide all of $v - u$ by a.

Example

Sometimes the subject appears in more than one term.

Make x the subject of the formula $a = \frac{x + c}{x - d}$.

$a = \frac{x + c}{x - d}$	Multiply both sides by $(x - d)$.
$a(x - d) = x + c$	Multiply out the brackets.
$ax - ad = x + c$	Collect like terms involving x on one side of the equation.
$ax - x = c + ad$	Factorise.
$x(a - 1) = c + ad$	
$x = \frac{c + ad}{(a - 1)}$	

These types of questions are usually worth at least 3 marks.

PROGRESS CHECK

1. Simplify the following expressions:

   a. $6a - 3b + 2a - 4b$

   b. $3a^2 - 6b^2 - 2b^2 + a^2$

   c. $5xy - 3yx + 2xy^2$

2.  If $a = \frac{3}{5}$ and $b = -2$, find the value of these expressions, giving your answer to 3sf where appropriate:

   a. $ab - 5$

   b. $a^2 + b^2$

   c. $3a - 6ab$

3. Make u the subject of the formula

   $v^2 = u^2 + 2as$.

4. Make p the subject of the formula

   $q = \frac{p - t}{p + v}$.

EXAM QUESTION

1. a. Sarah says 'when $x = 2$ the value of $3x^2$ is 36'.

   Josh says 'when $x = 2$ the value of $3x^2$ is 12'.

   Who is right? Explain why.

   b. Work out the value of $4(x - 1)^2$ when $x = 4$.

   c. If $y = 4(x - 1)^2$ make x the subject of the formula.

Brackets and Factorisation

Multiplying out brackets helps to simplify algebraic expressions.

Single brackets

Each term outside the brackets multiplies each separate term inside the bracket.

Examples

Expand and simplify:

a. $5(x + 6) = 5x + 30$

b. $-2(2x + 4) = -4x - 8$

c. $5(2x - 3) = 10x - 15$

d. $8(x + 3) + 2(x - 1)$

$= 8x + 24 + 2x - 2$

$= 10x + 22$ Collect like terms.

e. $3(2x - 5) - 2(x - 3)$

$= 6x - 15 - 2x + 6$ Multiply out the brackets.

$= 4x - 9$ Collect like terms.

Two brackets

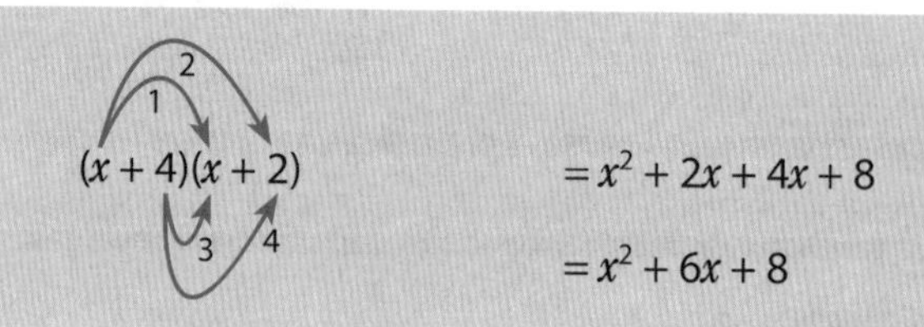

- Every term in the second bracket must be multiplied by every term in the first bracket.
- Often, but not always, the two middle terms are like terms and can be collected together.

Examples

$(x + 4)(2x - 5) = 2x^2 - 5x + 8x - 20$

$= 2x^2 + 3x - 20$

$(2x + 1)^2 = (2x + 1)(2x + 1)$

$= 4x^2 + 2x + 2x + 1$

$= 4x^2 + 4x + 1$

Remember that x^2 means x multiplied by itself.

$(3x - 1)(x - 2) = 3x^2 - 6x - x + 2$

$= 3x^2 - 7x + 2$

$(x - 4)(3x + 1) = 3x^2 + x - 12x - 4$

$= 3x^2 - 11x - 4$

Factorisation

Factorisation simply means putting into brackets.

One bracket

$4x + 6 = 2(2x + 3)$

To factorise $4x + 6$

- Recognise that 2 is the highest common factor of 4 and 6.
- Take out the common factor.
- The expression is completed inside the bracket so that when multiplied out it is equivalent to $4x + 6$.

Two brackets

Two brackets are obtained when a quadratic expression of the type $ax^2 + bx + c$ is factorised.

Examples

$x^2 + 4x + 3 = (x + 1)(x + 3)$

$x^2 - 7x + 12 = (x - 3)(x - 4)$

$x^2 + 3x - 10 = (x + 5)(x - 2)$

PROGRESS CHECK

1. Expand and simplify:
   a. $(x + 3)(x - 2)$
   b. $2(3x - 4)$
   c. $4x(x - 3)$
   d. $(x - 3)^2$
2. Factorise:
   a. $4x^2 + 8x$
   b. $12xy - 6x^2$
   c. $3a^2b + 6ab^2$
3. Factorise:
   a. $x^2 + 4x + 4$
   b. $x^2 - 5x + 6$
   c. $x^2 - 4x - 5$

EXAM QUESTION

1. a. Expand and simplify
   i. $t(3t - 4)$
   ii. $4(2x - 1) - 2(x - 4)$

   b. Factorise
   i. $y^2 + y$
   ii. $5p^2q - 10pq^2$
   iii. $(a + b)^2 + 4(a + b)$
   iv. $x^2 - 5x + 6$

Equations I

These involve an unknown value that needs to be worked out.

Linear equations

Type 1 of the form $ax + b = c$

Remember to do the same thing to both sides of the equation so that they balance.

Example

i. Solve $3x = 15$

$x = \frac{15}{3}$

$x = 5$

ii. Solve $\frac{x}{3} = 6$

$x = 6 \times 3$

$x = 18$

iii. Solve: $5x - 2 = 13$

$5x = 13 + 2$ — Add 2 to both sides.

$5x = 15$ — Divide both sides by 5.

$x = 15 \div 5$

$x = 3$

iv. Solve $3x + 1 = 13$

$3x = 13 - 1$

$3x = 12$

$x = \frac{12}{3}$

$x = 4$

v. Solve $\frac{x}{6} - 1 = 3$

$\frac{x}{6} = 3 + 1$

$\frac{x}{6} = 4$

$x = 4 \times 6$

$x = 24$

Type 2 of the form $ax + b = cx + d$

Example

i. Solve:

Add 4 to each side. Subtract $3x$ from both sides.

$7x - 4 = 3x + 8$

$7x = 3x + 12$

$4x = 12$

$x = 12 \div 4$

$x = 3$

Must check:

$7 \times 3 - 4 = 3 \times 3 + 8$

$21 - 4 = 9 + 8$

Yes, it's right. ✔

ii. Solve:

$5x + 3 = 2x - 5$

$5x = 2x - 5 - 3$

$5x = 2x - 8$

$5x - 2x = -8$

$3x = -8$

$x = -\frac{8}{3}$

$x = -2\frac{2}{3}$

Type 3 with brackets!

Examples

i. Solve: $5(x-1)=3(x+2)$

$$5x-5=3x+6$$
$$5x=3x+11$$
$$2x=11$$
$$x=11\div 2$$
$$x=5.5$$

Just multiply out the brackets and solve as normal.

ii. Solve: $5(2x+3)=2(x-6)$

$$10x+15=2x-12$$
$$10x=2x-12-15$$
$$10x=2x-27$$
$$10x-2x=-27$$
$$8x=-27$$
$$x=-\frac{27}{8}$$
$$x=-3\tfrac{3}{8}$$

iii. Solve: $\frac{3(2x-1)}{5}=6$

$$3(2x-1)=6\times 5$$
$$6x-3=30$$
$$6x=33$$
$$x=33\div 6$$
$$x=5.5$$

Multiply both sides by 5.

PROGRESS CHECK

Solve the following equations:

1. $2x-6=10$
2. $5-3x=20$
3. $4(2-2x)=12$
4. $6x+3=2x-10$
5. $7x-4=3x-6$
6. $5(x+1)=3(2x-4)$

EXAM QUESTION

1. Solve the equations.
   a. $5x-3=9$
   b. $7x+4=3x-6$
   c. $3(4y-1)=21$
2. Solve:
   a. $5-2x=3(x+2)$
   b. $\frac{3x-1}{3}=4+2x$

Equations 2

Equation problems

Always write down the information that you know.

Example

The perimeter of this rectangle is 30 cm. Work out the value of y and find the length of the rectangle.

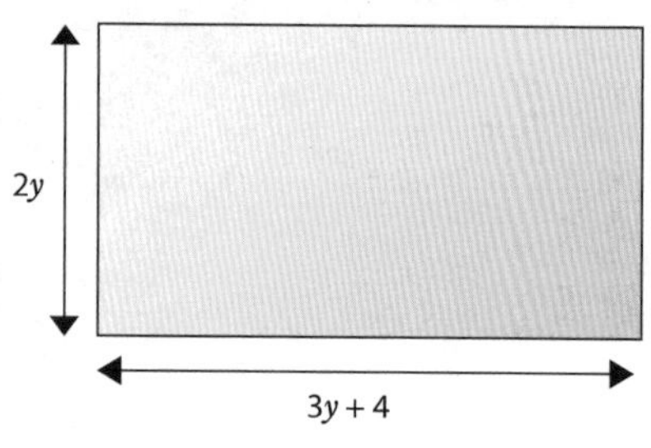

Write down what you know.

$3y + 4 + 2y + 3y + 4 + 2y = 30$

Simplify the expression and solve as normal.

$10y + 8 = 30$

$10y = 30 - 8$

$10y = 22$

$y = 2.2$

Length of rectangle $= 3 \times 2.2 + 4$

$= 10.6$ cm

Solving equations involving indices

Equations sometimes involve indices. You need to remember the laws of indices to be able to solve them.

Example

Solve:

1. $y^k = \sqrt[3]{y} \div \frac{1}{y^5}$

$y^k = y^{\frac{1}{3}} \div y^{-5}$ — Rewrite as indices.

$y^k = y^{\frac{1}{3} - (-5)}$ — When dividing subtract the indices.

$y^k = y^{5\frac{1}{3}}$ — Compare the indices since the bases are the same.

$k = 5\frac{1}{3}$

Example

Solve:

$2^{k+2} = 32$

$2^{k+2} = 2^5$ — Rewrite so that both bases are the same.

$k + 2 = 5$

$k = 3$

Solve:

$3^k = 27$

$3^k = 3^3$

$k = 3$

Example

The area of this rectangle is 81cm^2

3^{k+2}

$\sqrt{3}$

Work out the value of k.

$3^{k+2} \times \sqrt{3} = 81$	Write out the equation.
$3^{k+2} \times 3^{\frac{1}{2}} = 81$	
$3^{(k+2+\frac{1}{2})} = 3^4$	Write out so that the bases are 3.
$3^{(k+\frac{5}{2})} = 3^4$	Add the indices of the LHS.
$k + \frac{5}{2} = 4$	Compare the indices.
$k = \frac{3}{2}$	

Progress Check

1. The perimeter of this triangle is 60 cm. Work out the value of x and find the shortest length.

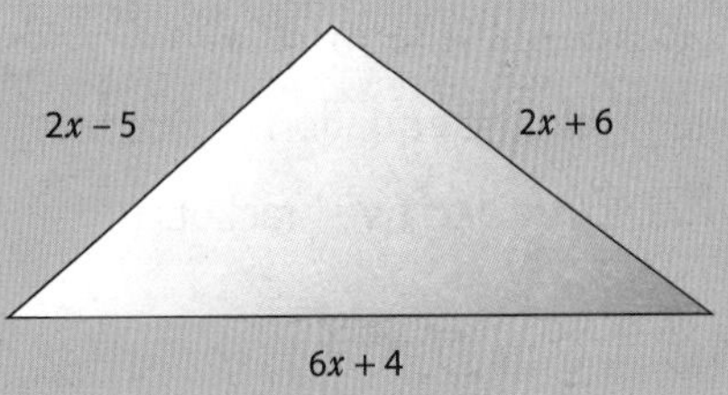

2. Solve $5^{k+3} = 5^{\frac{1}{3}}$
3. Solve $16^k = 64$
4. Solve $2^{3k-1} = 64$

Exam Question

1. The sizes of the angles, in degrees, of the quadrilateral are:

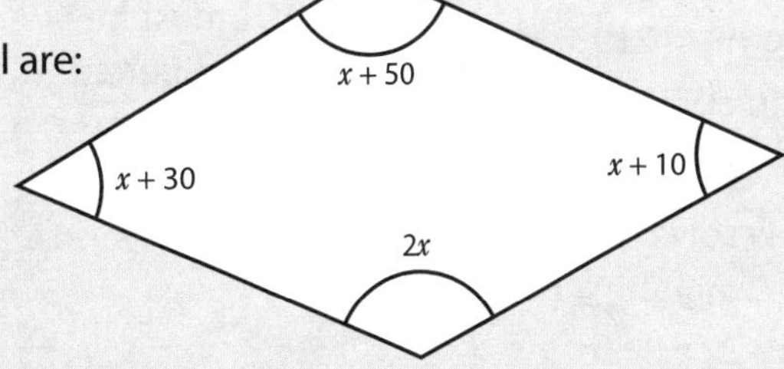

$x + 30$
$2x$
$x + 50$
$x + 10$

a. Use this information to write down an equation in terms of x.

b. Use your answer to part a. to work out the smallest angle of the quadrilateral.

Solving quadratic equations I

Quadratic equations can be solved using a variety of methods.

Solving quadratic equations

Solve: $x^2 - x - 6 = 0$

- It's a quadratic of the form $ax^2 + bx + c = 0$
- Must check that the equation equals zero.
- Need to factorise into two brackets $(\quad)(\quad) = 0$

Method 1: Factorisation

Example

Solve $2x^2 - x - 3 = 0$.

- Write out the two brackets and put an x and $2x$ in each one, since $x \times 2x = 2x^2$.
  $(2x \quad)(x \quad)$
- We now need two numbers that multiply to give –3 (one positive and one negative) and when multiplied by the x terms add up to give $-1x$.
  $(2x + 1)(x - 3)$
  $1 \times -3 = -3$ ✓
  x-terms $= -6x + x = -5x$ ✗ incorrect
  $(2x - 3)(x + 1)$
  $-3 \times 1 = -3$ ✓
  x-terms $= 2x - 3x = -x$ ✓ correct
  A quick check gives
  $(2x - 3)(x + 1) = 2x^2 - x - 3$

The $2x$ and 1 must be in different brackets.

- Now solve the equation:
  $(2x - 3)(x + 1) = 0$

$\therefore\ (2x - 3) = 0$ so $x = \frac{3}{2}$

or $(x + 1) = 0$ so $x = -1$

$\therefore\ x = \frac{3}{2}$ or -1

Method 2 The quadratic formula

Where a quadratic does not factorise then you can use the formula:

$$x = \frac{-b \pm \sqrt{b^2 - 4ac}}{2a}$$

for any quadratic equation written in the form $ax^2 + bx + c = 0$

This formula will be given on the exam paper.

PROGRESS CHECK

Solve:

1. a. $x^2 - 6x + 8 = 0$
   b. $x^2 + 5x + 4 = 0$
   c. $x^2 - 4x - 12 = 0$
2. Solve these equations by (i) factorisation (ii) using the quadratic formula.
   a. $x^2 + 2x - 15 = 0$
   b. $2x^2 + 5x + 2 = 0$

Example

Solve the equation $2x^2 - 7x = 5$.
Give your answers to 2 decimal places.

a. Put the equation into the form $ax^2 + bx + c = 0$: $2x^2 - 7x - 5 = 0$
b. Identify the values of a, b and c: $a = 2$, $b = -7$, $c = -5$
c. Substitute these values into the quadratic formula:

$$x = \frac{-b \pm \sqrt{b^2 - 4ac}}{2a}$$

$$x = \frac{7 \pm \sqrt{(-7)^2 - (4 \times 2 \times -5)}}{2 \times 2}$$

$$x = \frac{7 \pm \sqrt{49 - (-40)}}{4}$$

$$x = \frac{7 \pm \sqrt{89}}{4}$$

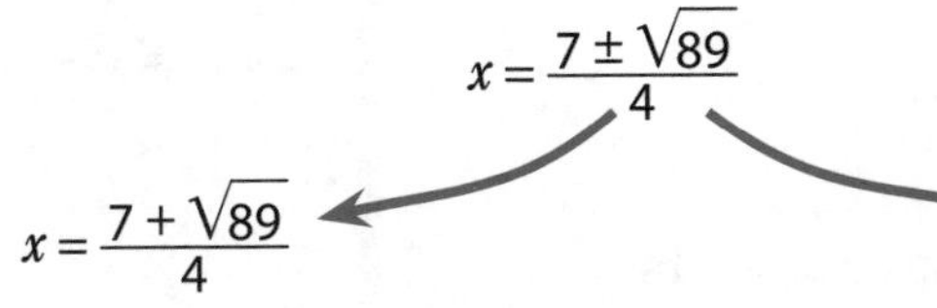

$$x = \frac{7 + \sqrt{89}}{4}$$

One solution is when we use $+\sqrt{89}$
$x = 4.11$ (2dp)

$$x = \frac{7 - \sqrt{89}}{4}$$

One solution is when we use $-\sqrt{89}$
$x = -0.61$ (2dp)

Check $2 \times (4.11)^2 - 7(4.11) - 5 = 0$ ✓ $2 \times (-0.61)^2 - 7(-0.61) - 5 = 0$ ✓

EXAM QUESTION

1. Charles cuts a square out of a rectangular piece of card.

   The length of the rectangle is $2x + 5$
   The width of the rectangle is $x + 3$
   The length of the side of the square is $x + 1$
   All measurements are in centimetres

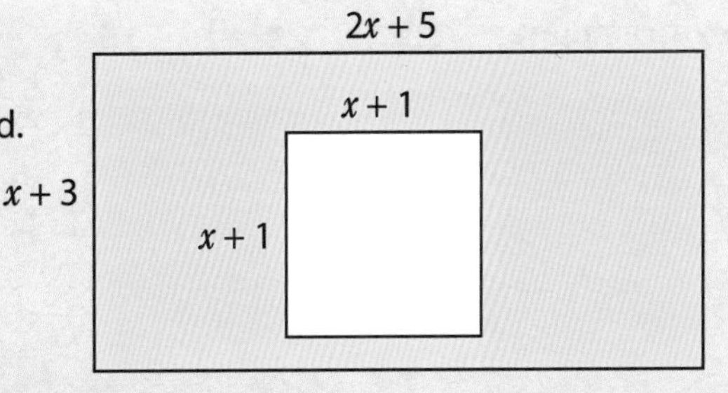

   The shaded shape in the diagram shows the card remaining.

   The area of the shaded shape is 45cm^2.

   a. Show that $x^2 + 9x - 31 = 0$

   b. i. Solve the equation $x^2 + 9x - 31 = 0$
   Give your answer correct to 3 significant figures.

   ii. Hence find the perimeter of the square.
   Give your answer correct to 3 significant figures.

Solving quadratic equations 2

Completing the square

This is when quadratic equations are expressed in the form

$$(x+p)^2+q=0$$

or

$$x^2+2px+p^2+q=0$$

1. Rearrange the equation into the form $ax^2+bx+c=0$
2. If a is not 1, divide the whole equation by a.
3. Write the equation in the form $(x+\frac{b}{2})^2$.

   Notice that the number in the bracket is always half the value of b.
4. Multiply out the brackets, compare to the original and adjust by adding or subtracting an extra amount.

Example

Express $x^2-6x+2=0$ as a completed square and hence solve it.

$x^2-6x+2=0$	Rearrange in the form $ax^2+bx+c=0$.
$(x-3)^2$	Half of -6 is -3.
$(x-3)^2=x^2-6x+9$	Multiply out the brackets and now compare to the original. x^2-6x+2
$(x-3)^2-7=0$	To make the expression equal, we subtract 7. $(x-3)^2-7=x^2-6x+2$ $\therefore\ (x-3)^2-7=0$

Now solve:

$$(x-3)^2=7$$

$$(x-3)=\pm\sqrt{7}$$

$$\therefore\quad x=\sqrt{7}+3 \text{ or } x=-\sqrt{7}+3$$

$$x=5.65 \text{ or } x=0.35$$

Solving cubic equations by trial and improvement

Trial and improvement gives an approximate solution to cubic equations.

Example

The equation $x^3 + 2x = 58$ has a solution between 3 and 4. Find the solution to 1 decimal place.

Drawing a table can help you – but also the examiner – since it makes it easier to follow what you have done.

x	$x^3 + 2x$	Comment
3.5	$3.5^3 + 2 \times 3.5 = 49.875$	too small
3.8	$3.8^3 + 2 \times 3.8 = 62.472$	too big
3.7	$3.7^3 + 2 \times 3.7 = 58.053$	too big
3.65	$3.65^3 + 2 \times 3.65 = 55.927$	too small

$x = 3.7$ (1dp)

1. The equation $a^3 = 40 - a$ has a solution between 3 and 4. Find the solution to 1dp, by using a method of trial and improvement.

2. The expression $x^2 + 8x + 2$ can be written in the form $(x + a)^2 + b$ for all values of x.

   a. Find a and b.

   b. The expression $x^2 + 8x + 2$ has a minimum value. Find this minimum value.

EXAM QUESTION

1. The equation
   $x^3 + 4x^2 = 49$
   has a solution between $x = 2$ and $x = 3$.

   Use a trial and improvement method to find this solution. Give your answer correct to 1 decimal place. You must show all your working.

2. The expression $x^2 - 4x + 7$ can be written in the form $(x + p)^2 + q$ for all values of x.
   Find the values of p and q.

Simultaneous linear equations

Two equations with two unknowns are called simultaneous equations. They can be solved graphically or algebraically.

By algebra (elimination method)

Solve simultaneously:

$$3x + 2y = 8$$
$$2x - 3y = 14$$

Label the equations ① and ②.

$$3x + 2y = 8 \quad ①$$
$$2x - 3y = 14 \quad ②$$

Since no coefficients match, multiply equation ① by 2 and equation ② by 3.

$$6x + 4y = 16$$
$$6x - 9y = 42$$

Rename them equations ③ and ④.

$$6x + 4y = 16 \quad ③$$
$$6x - 9y = 42 \quad ④$$

The coefficient of x in equations ③ and ④ is the same. Subtract equation ④ from equation ③ and solve the remaining equation.

$$x + 13y = -26$$
$$y = -26 \div 13$$
$$y = -2$$

Substitute the value of $y = -2$ back into equation ①. Solve this equation to find x.

$$3x + (-4) = 8$$
$$3x = 8 + 4$$
$$3x = 12$$
$$x = 4$$

Check in equation ②.

$$(2 \times 4) - (3 \times -2) = 14 \checkmark$$

Solution is: $x = 4, y = -2$

Graphically

Solve the simultaneous equations.

$2x + 3y = 6$
$x + y = 1$

The point at which any two graphs intersect represents the simultaneous solutions of their equations.

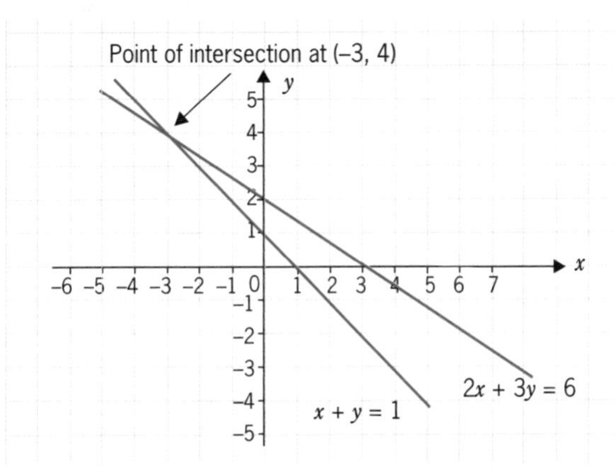

Draw the graph of:

$2x + 3y = 6$

$x = 0 \; 3y = 6 \; \therefore \; y = 2 \; (0, 2)$

$y = 0 \; 2x = 6 \; \therefore \; x = 3 \; (3, 0)$

Draw the graph of:

$x + y = 1$

$x = 0, y = 1 \; \therefore \; (0, 1)$

$y = 0, x = 1 \; \therefore \; (1, 0)$

At the point of intersection: $x = -3, y = 4$

Progress Check

1. Solve the following pairs of simultaneous equations:

   a. $4b + 7a = 10$
      $2b + 3a = 3$

   b. $2p + 3r = 6$
      $p + r = 1$

   c. $y - 2x = -1$
      $x + y = 5$

2. The diagram shows the graphs of the equations:

   $x + y = 6$ and $y = x + 2$

   Use the diagram to solve the simultaneous equations $x + y = 6$ and $y = x + 2$

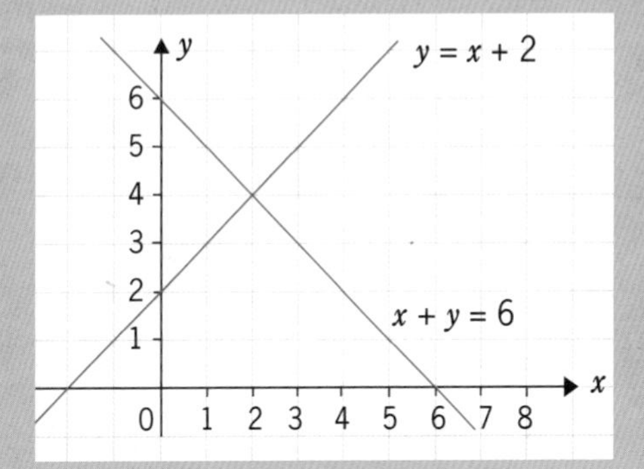

Exam Question

1. Solve the simultaneous equations.

   a. $5a - 2b = 19$

   b. $3a + 4b = 1$

Solving linear and quadratic equations simultaneously

You need to be able to work out the coordinates of the points of intersection of a straight line and a quadratic curve as well as a straight line and a circle.

Point of intersection of a straight line and a curved graph

The diagram shows the points of intersection A and B of the line $y = 4x - 2$ and the curve $y = x^2 + 1$. We know that the points A and B lie on both the curve and the line.

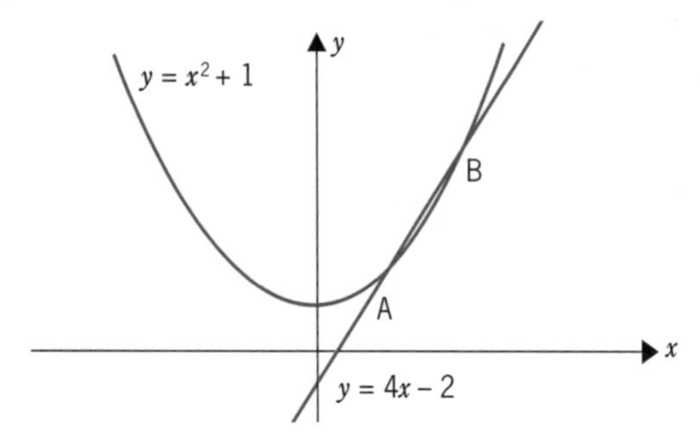

$y = 4x - 2$ ①
$y = x^2 + 1$ ②

Eliminate y by substituting equation ② into equation ①.

$x^2 + 1 = 4x - 2$
rearrange $x^2 - 4x + 3 = 0$
factorise $(x - 1)(x - 3) = 0$
so $x = 1$ and $x = 3$

Now substitute the values of x back into equation ② to find the corresponding values of y.

when $x = 1, y = 1^2 + 1 = 2$
when $x = 3, y = 3^2 + 1 = 10$

Coordinates of A are (1, 2). Coordinates of B are (3, 10).

Points of intersection of a straight line and a circle

The equation of a circle with centre (0, 0) and radius r is $x^2 + y^2 = r^2$.

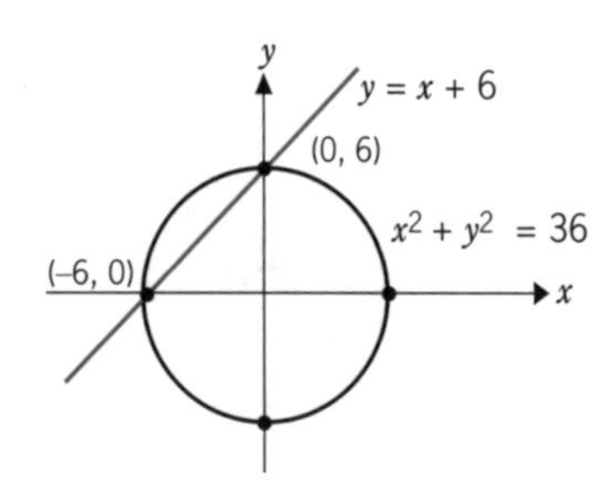

Find the coordinates of the points where the line $y - x = 6$ cuts $x^2 + y^2 = 36$.

The coordinates must satisfy both equations so we solve them simultaneously.

$y - x = 6$ ①
$x^2 + y^2 = 36$ ②

Rewrite equation ① in the form '$y =$'.

$y = x + 6$ ①

Eliminate y by substituting equation ① into equation ②

$x^2 + (x+6)^2 = 36$ — Expand brackets.

$$(x+6)^2 = (x+6)(x+6)$$
$$= x^2 + 12x + 36$$

$x^2 + x^2 + 12x + 36 = 36$

$2x^2 + 12x = 0$ — Rearrange.

$2x(x+6) = 0$ — Factorise.

$x = 0, x = -6$ — Solve.

Substitute the values of x into equation ① to find the values of y.

$x = 0, y = 0 + 6 = 6$

$x = -6, y = -6 + 6 = 0$

The line $y - x = 6$ cuts the circle $x^2 + y^2 = 36$ at (0, 6) and (–6, 0).

1. Solve these simultaneous equations

   a. $y = x + 4$

   $y = x^2 + 2$

   b. $y = x + 1$

   $x^2 + y^2 = 25$

2. For the two questions above give a geometrical interpretation of the result.

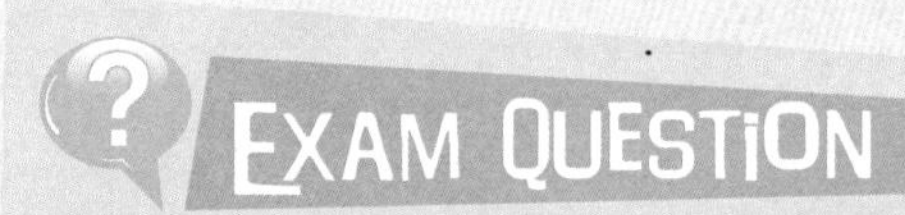

1. Solve the simultaneous equations.

   a. $x^2 + y^2 = 26$ $\qquad x =, y =$

   b. $y = 2x + 3$ $\qquad$ and $x =, y =$

Algebraic fractions

When working with algebraic fractions, the same rules are used as with ordinary fractions.

Simplifying algebraic fractions

Example

Simplify $\frac{8x + 16}{x^2 - 4}$.

- Firstly you need to factorise the numerator and denominator.

  $\frac{8(x + 2)}{(x + 2)(x - 2)}$

  Remember the difference of two squares: $(b^2 - a^2) = (b - a)(b + a)$

- Now cancel any common factors, i.e. the $(x + 2)$.

  $\frac{8\cancel{(x + 2)}}{\cancel{(x + 2)}(x - 2)} = \frac{8}{(x - 2)}$

Multiplication

- Multiply the numerators together and the denominators together.
- Cancel if possible.

Example

$\frac{3a^2}{4b} \times \frac{16b^2}{9ab} = \frac{48a^2b^2}{36ab^2} = \frac{4}{3}a$

Division

- Remember to turn the second fraction upside down (i.e. take the reciprocal) then multiply and cancel if possible.

Example

$\frac{12(x - 3)}{(x + 2)} \div \frac{4(x - 1)(x - 3)}{(x + 1)}$

Cancel out common factors

$= \frac{{}^{3}\cancel{12}\cancel{(x - 3)}}{(x + 2)} \times \frac{(x + 1)}{\cancel{4}(x - 1)\cancel{(x - 3)}}$

$= \frac{3(x + 1)}{(x - 1)(x + 2)}$

Addition and subtraction

- As with ordinary fractions, you cannot add or subtract unless the fractions have the same denominator.

Example

$$\frac{(x+3)}{(x+2)} + \frac{3}{(x-1)}$$

1. The common denominator is: $(x+2)(x-1)$.

2. The numerator now needs to be adjusted:

$$\frac{(x+3)(x-1)+3(x+2)}{(x+2)(x-1)}$$

3. Multiply out the numerator:

$$\frac{x^2+2x-3+3x+6}{(x+2)(x-1)}$$

4. Simplify, then factorise the numerator if possible and then simplify by cancelling:

$$=\frac{x^2+5x+3}{(x+2)(x-1)}$$

PROGRESS CHECK

1. Decide whether these expressions are fully simplified:

a. $\frac{6s^2w}{3w^2}$

b. $\frac{5a^2b}{2c}$

c. $\frac{9abc}{d}$

2. Simplify the following algebraic fractions:

a. $\frac{3}{(x+1)} + \frac{2}{(x-1)}$

b. $\frac{3a^2b}{2q} \times \frac{4q^2}{9abc}$

c. $\frac{10(a^2-b^2)}{3x+6} \div \frac{(a-b)}{4x+8}$

EXAM QUESTION

1. Simplify fully: $\frac{8x+16}{x^2-4}$

2. Simplify fully: $\frac{x^2(6+x)}{x^2-36}$

3. Simplify fully: $\frac{2x^2+7x-15}{x^2+3x-10}$

Sequences

A sequence is a set of numbers that follow a particular rule. The word 'term' is often used to describe the numbers in the sequence.

Special sequences

Odd numbers	1, 3, 5, 7, 9 …	nth term is $2n - 1$
Even numbers	2, 4, 6, 8, 10 …	nth term is $2n$

Square numbers

1 4 9 16 25 …

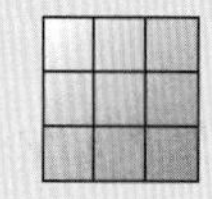

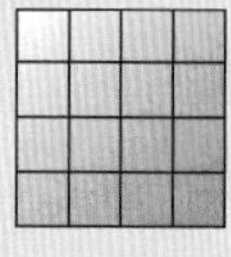

$1^2 = 1 \times 1$ $2^2 = 2 \times 2$ $3^2 = 3 \times 3$ $4^2 = 4 \times 4$ $5^2 = 5 \times 5$

Cube numbers

1 8 27 64 125 …

$1^3 = 1 \times 1 \times 1$ $2^3 = 2 \times 2 \times 2$ $3^3 = 3 \times 3 \times 3$

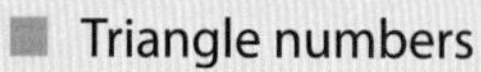

Triangle numbers

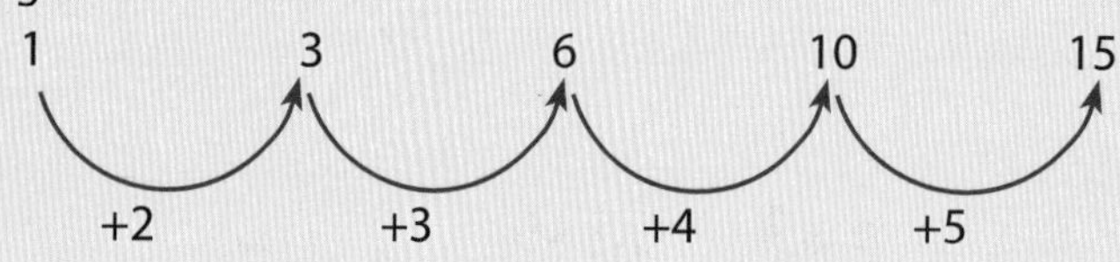

Fibonacci sequence

1, 1, 2, 3, 5, 8, 13 … Add the previous two terms.

Finding the nth term of a linear sequence

The nth term is often denoted by U_n. For example the 8th term is U_8.
For a linear sequence the nth term takes the form: $U_n = an + b$

Example

Find the nth term of this sequence:

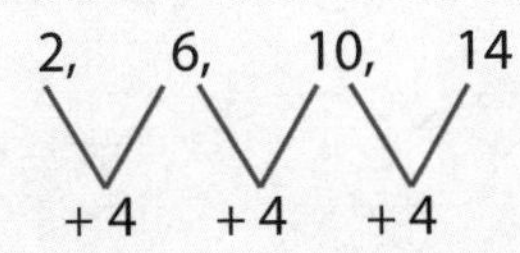

- See how the numbers are jumping (going up in 4s).
- The nth term is $4n$ + or – something.
- Try out $4n$ on the first term. This gives $4 \times 1 = 4$, but the first term is 2 … so subtract 2.
- The rule is $4n - 2$.
- Test this rule on the other terms

  $1 \rightarrow 4 - 2 = 2$

  $2 \rightarrow 8 - 2 = 6$

  $3 \rightarrow 12 - 2 = 10$

  It works on all of them.
- nth term is $4n - 2$
- The 20th term in the sequence would be: $4 \times 20 - 2 = 78$

PROGRESS CHECK

1. The cards show the nth term of some sequences:

$2n$	$4n + 1$	$3n + 2$	$5n - 1$	$2 - n$

   Match the cards with the sequences below:

   a. 5, 9, 13, 17, …

   b. 1, 0, –1, –2, …

   c. 2, 4, 6, 8, 10, …

   d. 5, 8, 11, 14, 17, …

   e. 4, 9, 14, 19, …

2. Find the nth term of these sequences:

   a. 7, 10, 13, 16, 19, …

   b. $\frac{1}{3}, \frac{1}{5}, \frac{1}{7}, \frac{1}{9}, \ldots$

   c. 1, 4, 7, 10 …

EXAM QUESTION

1. Here are the first four terms of an arithmetic sequence.

   5 7 9 11

   Find an expression in terms of n for the nth term of the sequence.

Inequalities

In inequalities the LHS does not equal the RHS.

Inequalities can be solved in exactly the same way as equations except that when multiplying or dividing by a negative number, you must reverse the inequality sign.

The inequality symbols

$>$ means **greater than**

$<$ means **less than**

$\geqslant$ means **greater than or equal to**

$\leqslant$ means **less than or equal to**

Example

Solve:

$2x - 2 < 10$

$2x < 10 + 2$

$2x < 12$

$x < 6$

Solve:

$3 - 2x \geqslant 9$

$-2x \geqslant 9 - 3$

$-2x \geqslant 6$

$x \leqslant \frac{6}{-2}$

$x \leqslant -3$

Divide by −2 and change inequality sign round.

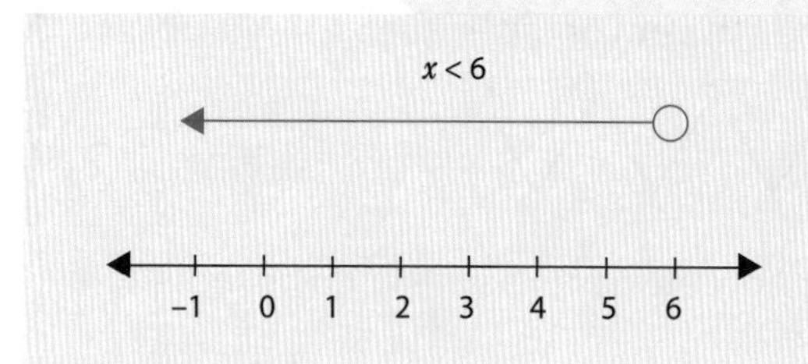

The open circle means that 6 is not included.

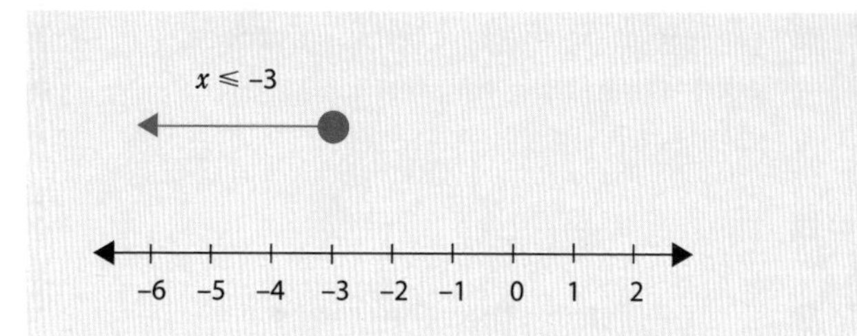

The solid circle means that −3 is included.

Example

Solve:

$-2 < 4x - 3 \leqslant 9$

$1 < 4x \leqslant 12$

$\frac{1}{4} < x \leqslant 3$

The integer values that satisfy this inequality are 1, 2, 3.

Graphs of inequalities

The graph of an equation such as $x = 2$ is a line, whereas the graph of the inequality $x < 2$ is a region that has $x = 2$ as its boundary.

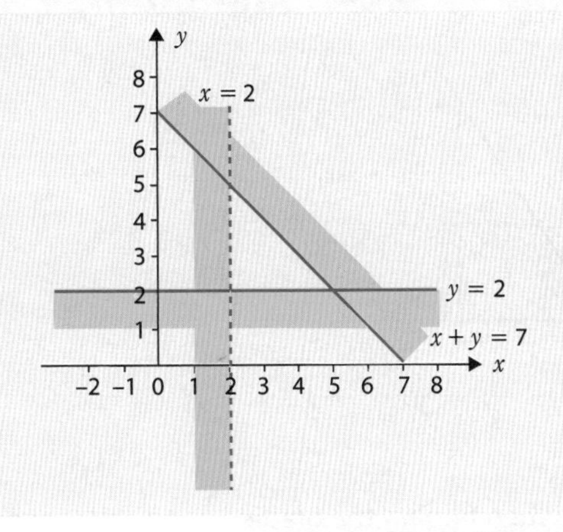

The diagram shows unshaded the region:

$x + y \leqslant 7$

$x > 2$

$y \geqslant 2$

For strict inequalities $>$ and $<$ the boundary line is not included and is shown as a dashed line.

PROGRESS CHECK

1. Solve the following inequalities:

   a. $5x - 1 < 10$

   b. $6 \leqslant 3x + 2 < 11$

   c. $3 - 5x < 12$

2. On the diagram below, leave unshaded the region satisfied by these inequalities:

   $x + y \leqslant 5$
   $x \geqslant 1$
   $y > 1$

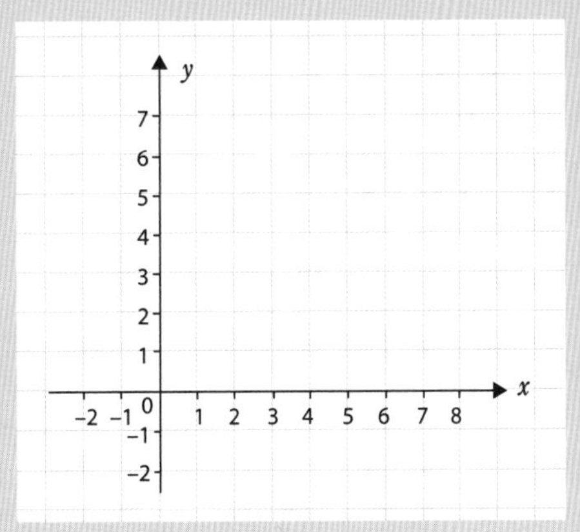

EXAM QUESTION

1. n is an integer such that $-6 < 2n \leqslant 8$

   a. List all the possible values of n

   b. Solve the inequality $4 + x > 7x - 8$

Straight-line graphs

The general equation of a straight line graph is

$$y = mx + c$$

m is the gradient. c is the intercept on the y-axis.

1. To work out the coordinates of the points that lie on the line $y = 3x - 4$, draw a table of values.

x	−1	0	2	4
y	−7	−4	2	8

2. Substitute the x values into the equation $y = 3x - 4$, to find the values of y
   e.g. $x = 2, y = 3 \times 2 - 4 = 2$

3. The coordinates of the points on the line are:

   (−1, −7) (0, −4)
   (2, 2) (4, 8)

   Just read them from the table of values.

4. Join the points with a straight line.

5. The graph $y = 3x - 4$ is drawn.

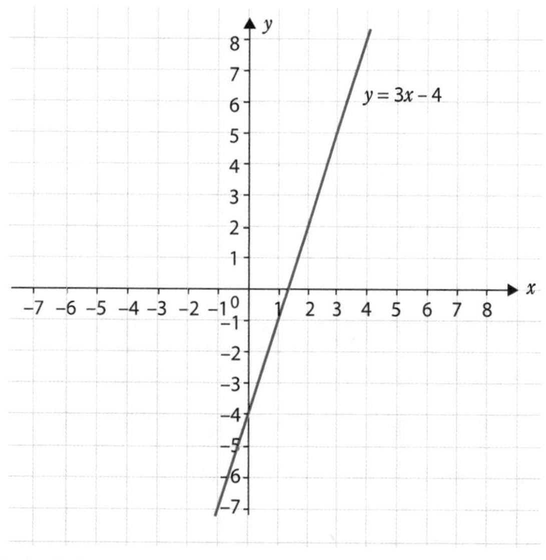

6. Label the graph once you've drawn it.

Gradient of a straight line

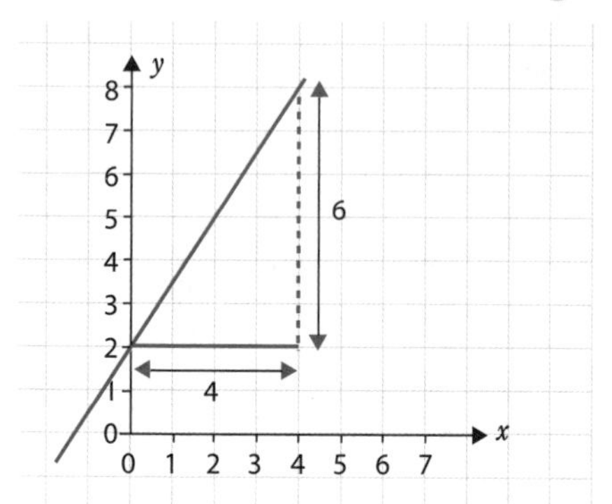

Be careful when finding the gradient – double-check the scales.

$$\text{Gradient} = \frac{\text{change in } y}{\text{change in } x} \text{ OR } \frac{\text{height}}{\text{base}}$$

$$\text{Gradient} = \frac{6}{4} = \frac{3}{2} \text{ or } 1.5$$

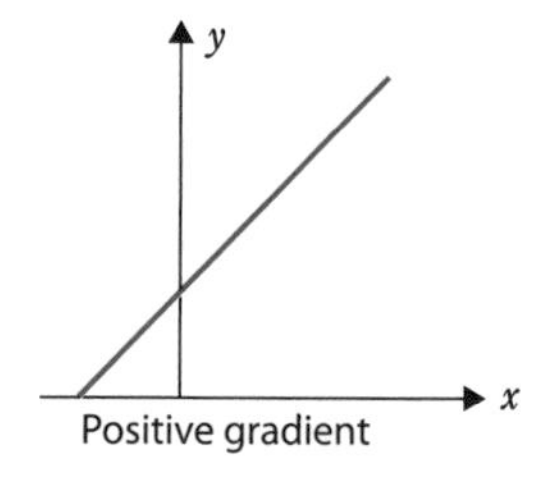

Positive gradient

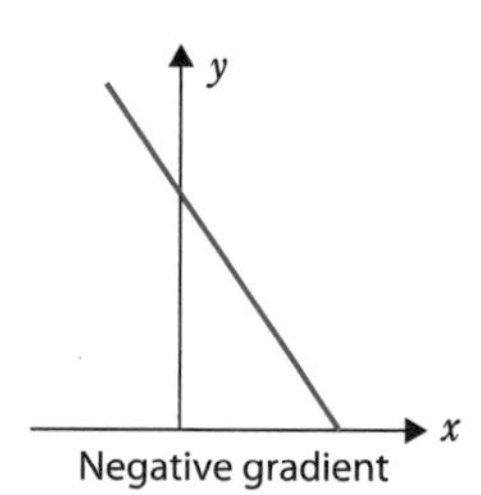

Negative gradient

Perpendicular lines

If two lines are **perpendicular** the product of their gradients is −1.

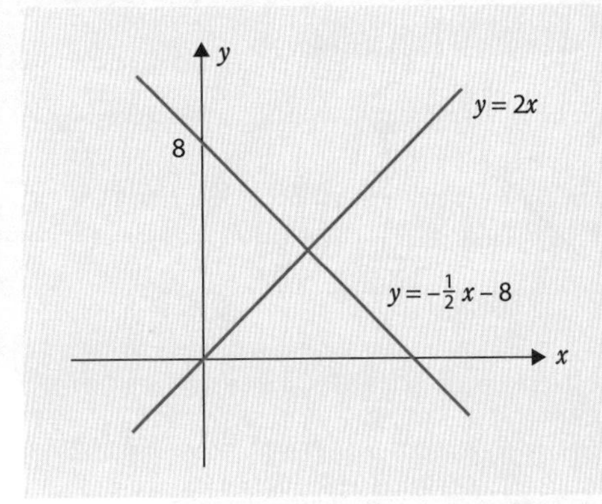

The lines $y = 2x$ and $y = -\frac{1}{2}x + 8$ are perpendicular because the gradients multiply to give −1

$(2 \times -\frac{1}{2} = -1)$

PROGRESS CHECK

1. a. Complete the table of values for $y = 2x + 3$.

x	−2	−1	0	1	2	3
y						

b. Draw the graph of $y = 2x + 3$

2. Write down the equation of the line *L*.

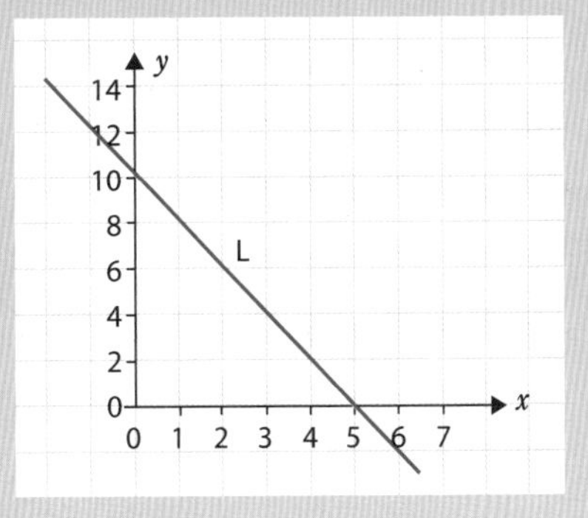

EXAM QUESTION

1. The diagram shows three points

   E (−2, 6) F (3, −4) G (0, 3)

   A line L is parallel to EF and passes through G.

   a. Find an equation for the line L

   b. A line K is perpendicular to EF and also passes through G.
   Find the equation of the line K.

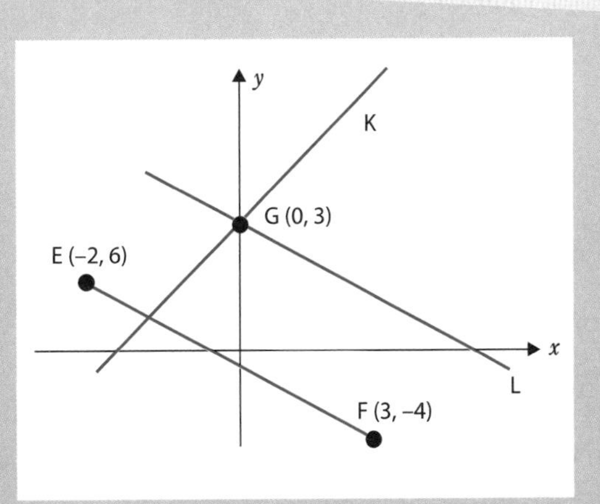

Curved graphs

Quadratic graphs have an x^2 term.
They will be ∪ shaped if the coefficient of x^2 is positive, and ∩ shaped if the coefficient of x^2 is negative.

Draw the graph of $y = x^2 - 2x - 6$.
Use values of x from −2 to 3.

1. Draw a table of values.

x	−2	−1	0	1	2	3
y	2	−3	−6	−7	−6	−3

2. Fill in the table of values by substituting the values of x into the equation.

   e.g. $x = 1$ $y = 1^2 - 2 \times 1 - 6$, $y = -7$
   Coordinates are (1, −7)

3. Draw the axes on graph paper and plot the points.

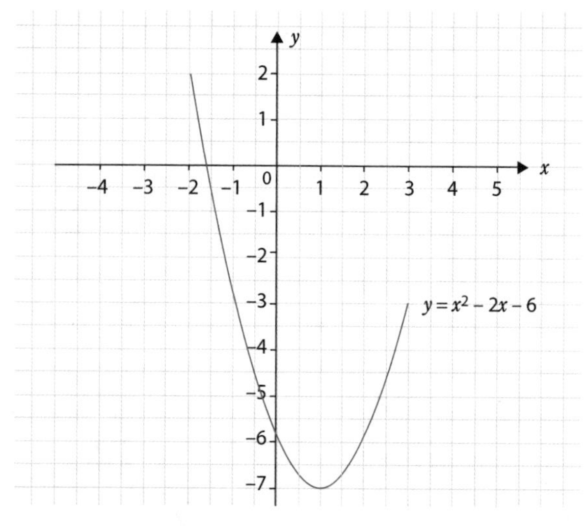

4. Join the points with a smooth curve.

5. Label the curve.

The minimum value is when $x = 1$, $y = -7$

The line of symmetry is at $x = 1$

Other graph shapes

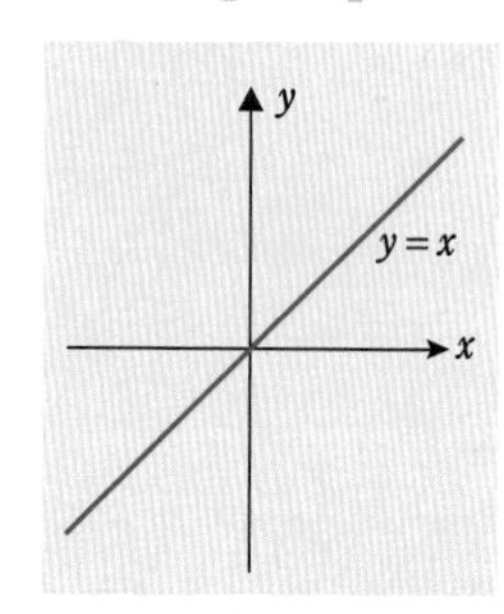

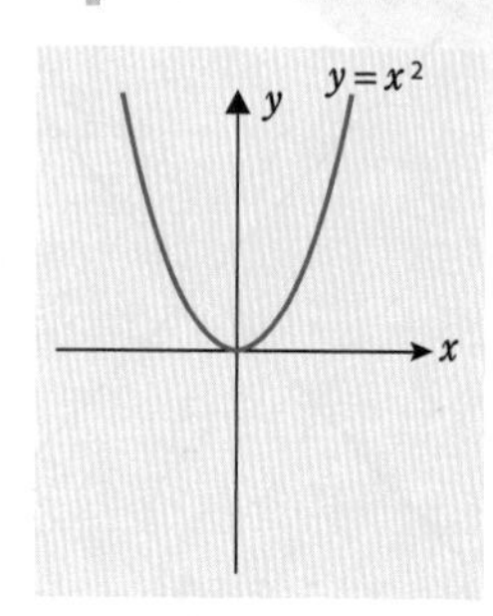

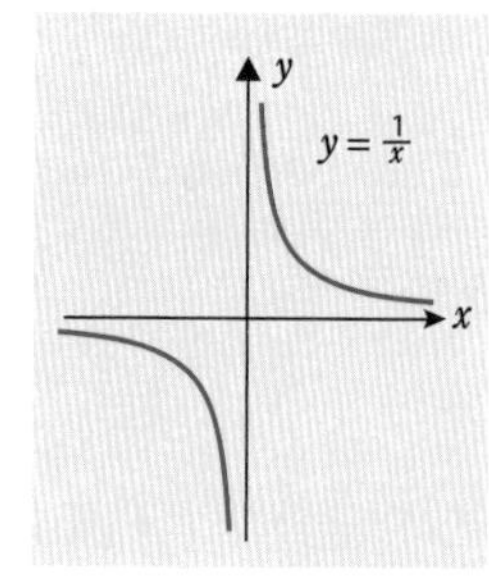

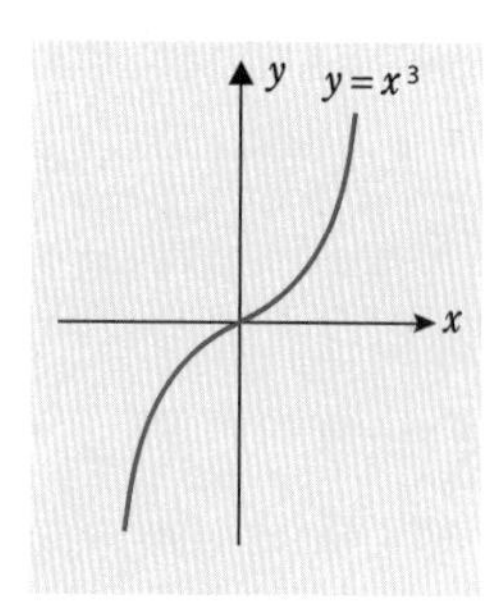

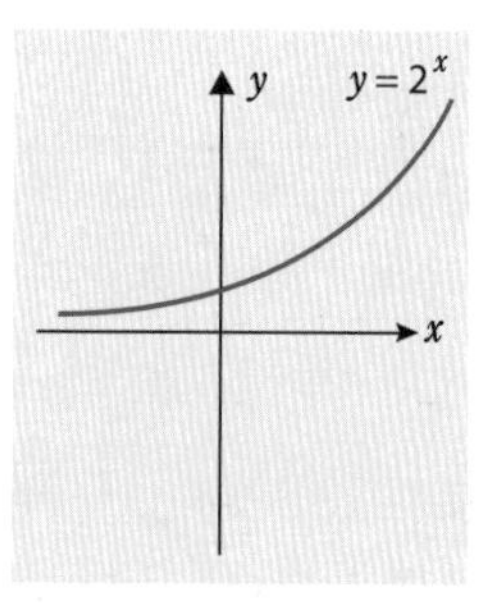

Example

Match each graph to the following equations.

Equation	Graph
$y = x^2 - 4$	C
$y = 5 - 2x$	D
$y = 2x^3 + 1$	A
$y = 3x - 1$	B

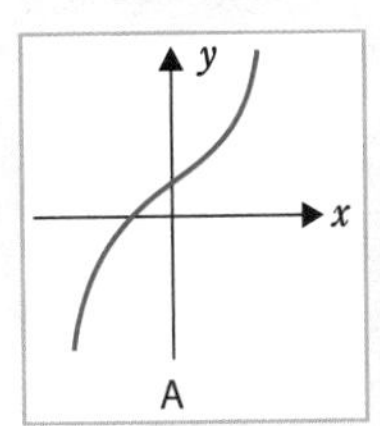

A

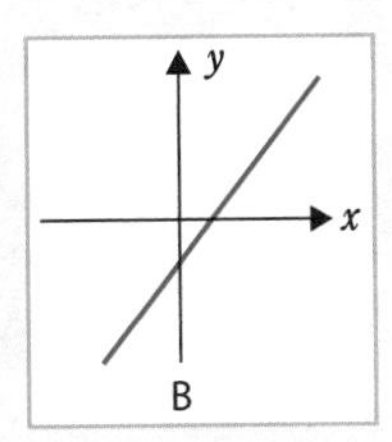

B

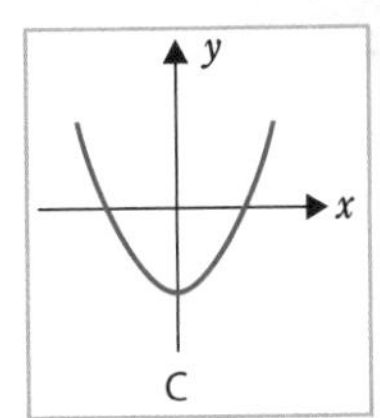

C

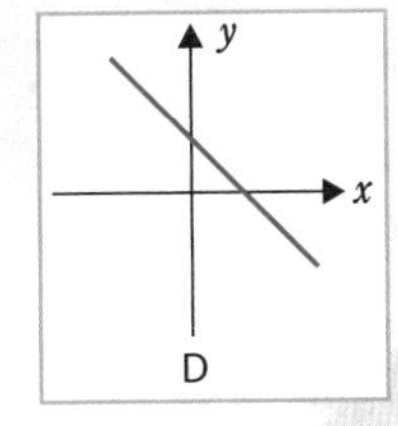

D

PROGRESS CHECK

1. a. Complete the table of values for $y = x^3 - 1$.

x	-3	-2	-1	0	1	2	3
y				-1			

b. Draw the graph of $y = x^3 - 1$.
Use scales of 2 units for 1 cm on the x-axis and 20 units for 1 cm on the y-axis.

c. From the graph find the value of x when $y = 15$.

EXAM QUESTION

1. Match each graph below to one of the equations.

$y = x^3 - 5$

$y = 2 - x^2$

$y = 4x + 2$

$y = \frac{3}{x}$

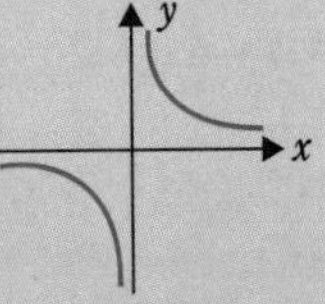

Graph A

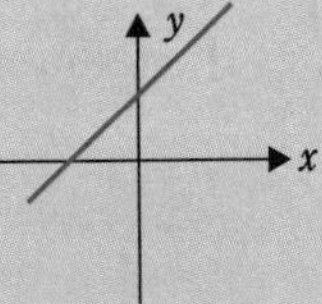

Graph B

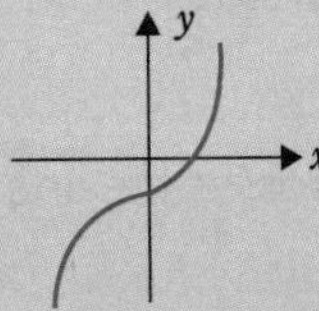

Graph C

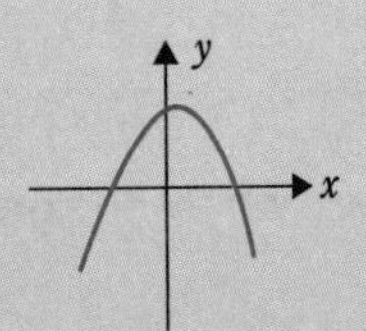

Graph D

Interpreting graphs

Accurate graphs can provide solutions to some equations.

Using graphs to solve equations

Often an equation needs to be rearranged in order to resemble the equation of the plotted graph.

The graph $y = x^2 - 4x + 4$ is drawn here:

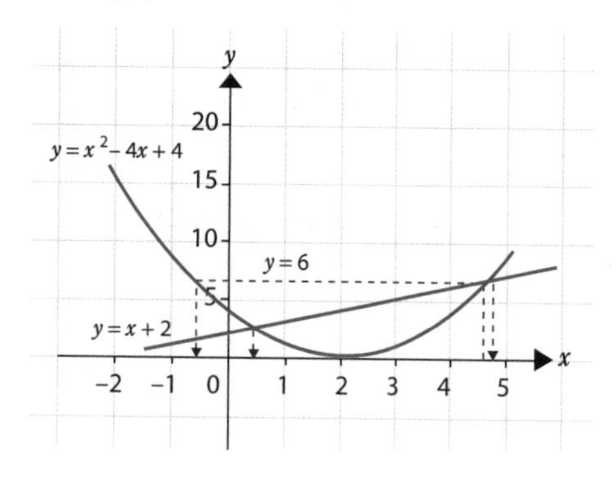

Use the graph to solve:

i. $x^2 - 4x + 4 = 6$ ii. $x^2 - 5x + 2 = 0$

For part (i) the equation needs to be arranged so it is the same as the plotted graph $y = x^2 - 4x + 4$

$x^2 - 4x + 4 = 6$

So the solution will be when the graph $y = x^2 - 4x + 4$ meets the line $y = 6$

$x = -0.4$ $x = 4.4$ (approximately)

For part (ii) we need to rearrange the equation:

$x^2 - 5x + 2 = 0$

$x^2 - 5x + 2 + x + 2 = x + 2$ — add $x + 2$ to both sides

$x^2 - 4x + 4 = x + 2$

The solutions are where the graph crosses the line $y = x + 2$

$x = 0.4$ $x = 4.6$

Using graphs to find relationships

This graph is known to fit $y = pq^x$.

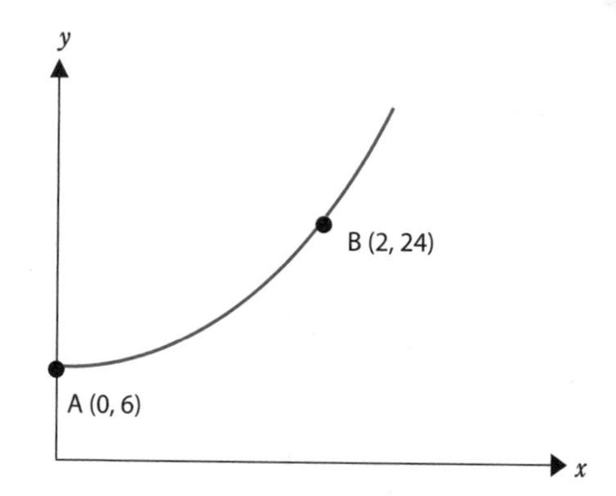

Use the graph to find the values of p and q and the relationship.

1. Take the value at point A and substitute into $y = pq^x$

   $6 = p \times q^0$

   Since $q^0 = 1$ $6 = p$

2. Take the coordinates at point B and substitute into $y = pq^x$

   $24 = 6 \times q^2$

   $\therefore$ $4 = q^2$

   $q = 2$

   The relationship is $y = 6 \times 2^x$

Speed–time graphs

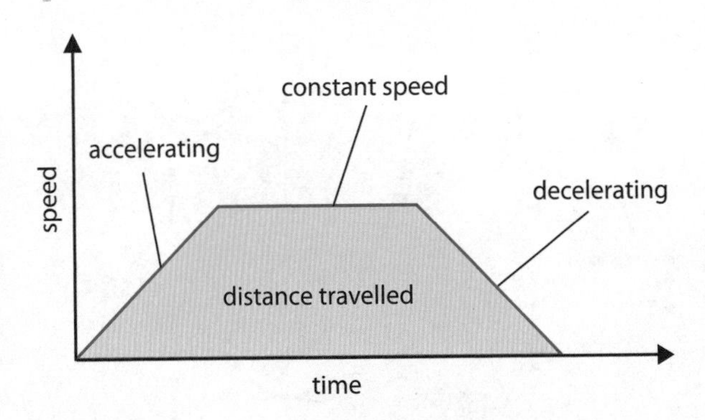

- Distance is the area between the graph and x-axis.
- A positive gradient means the speed is increasing.
- A negative gradient means the speed is decreasing.
- A horizontal line means the speed is constant.

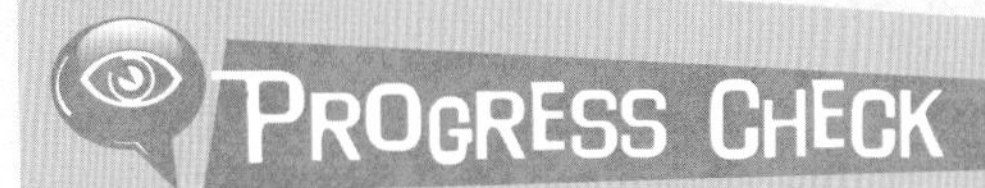

PROGRESS CHECK

Draw the graph of $y = x^2 + x - 2$.
Use your graph to decide whether the solutions of these equations are true or false.

1. $x^2 + x = 5$, solution is approximately $x = -2.8, x = 1.8$
2. $x^2 - 2 = 0$, solution is approximately $x = 2.4, x = -2.4$
3. $x^2 - 2x - 2 = 0$, solution is approximately $x = -0.7, x = 5.6$

EXAM QUESTION

1. Miss Jones has a car.

   The value of the car on September 1st 2004 was £9000
   The value of the car on September 1st 2006 was £4000

   The sketch graph shows how the value, £v, of the car changes with time. The equation of the sketch graph is:

   $V = ab^t$

   where t is the number of years after September 1st 2004.

   Use the information on the graph to find the value of a and b.

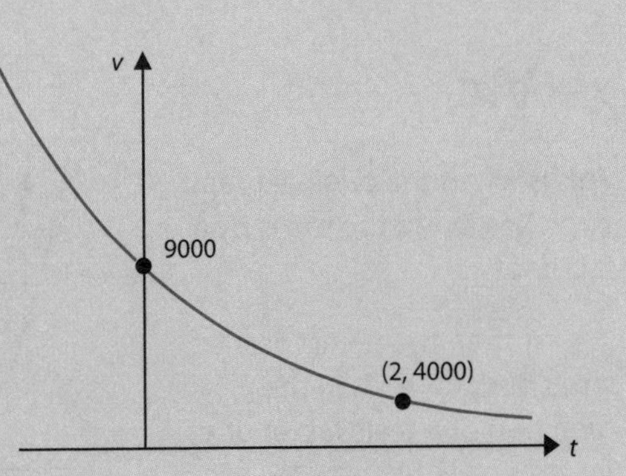

Functions and Transformations

If y = an expression involving x then it can be written as $y = f(x)$. The graphs of the related functions can be found by applying transformations.

$y = f(x) \pm a$

This is when the graphs move up or down the y-axis by a value of a.

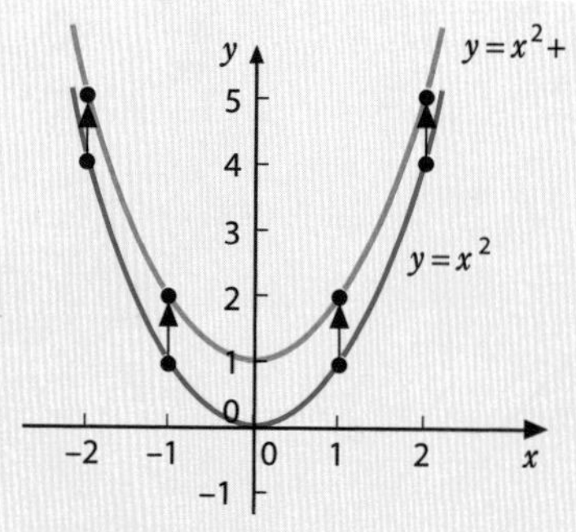

Graph moves up the y-axis by one unit.

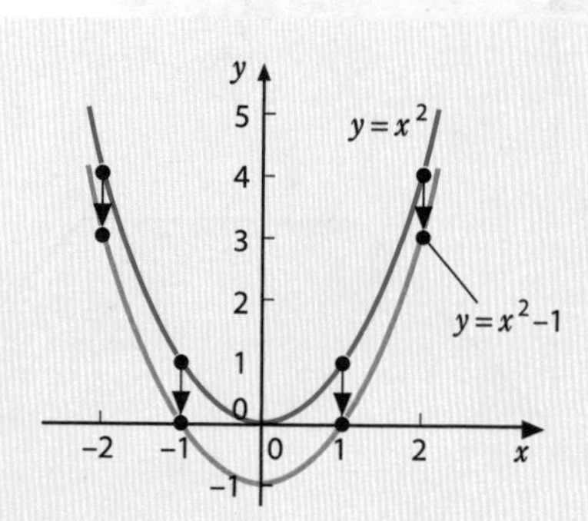

Graph moves down the y-axis by one unit.

$y = f(x \pm a)$

This is when the graphs move along the x-axis by a units.

$y = f(x + a)$ moves the graph a units to the left

$y = f(x - a)$ moves the graph a units to the right

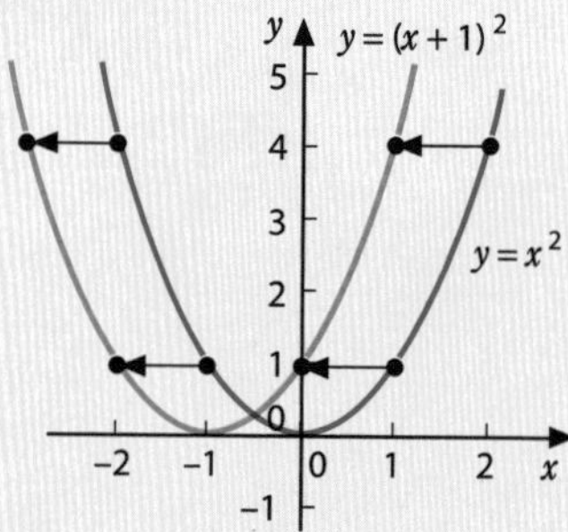

Graph of $y = (x + 1)^2$ moves one unit to the left.

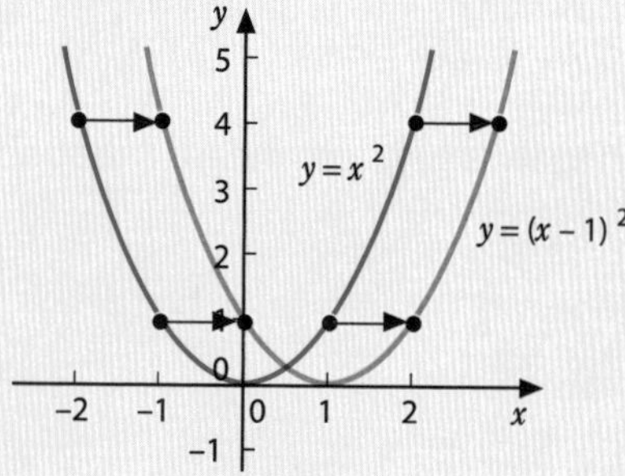

Graph of $y = (x - 1)^2$ moves one unit to the right.

$y = kf(x)$

This is when the original graph stretches along the y-axis by a factor of k.

If $k > 1$ then the points are stretched upwards in the y direction of a scale factor of k.

If $k < 1$, e.g. $y = \frac{1}{2}x^2$, the graph squashes downwards by a scale factor of $\frac{1}{2}$.

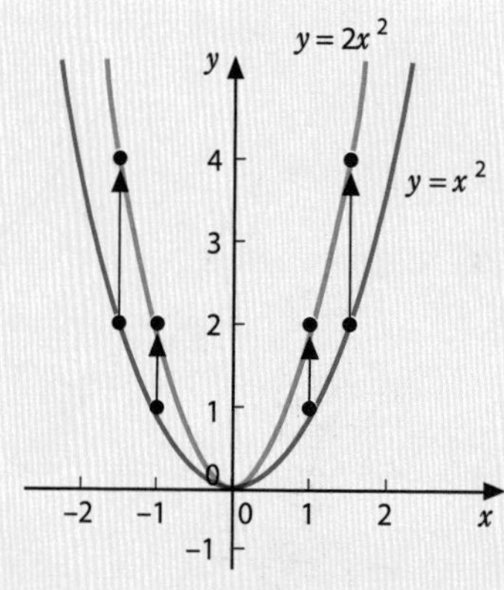

For this graph the x values stay the same and the y values are multiplied by 2.

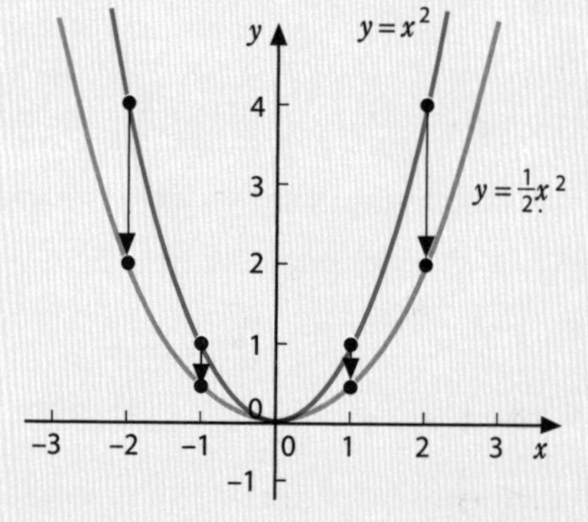

For this graph the x values stay the same and the y values are multiplied by $\frac{1}{2}$.

$y = f(kx)$

If $k > 1$ then the graph stretches inwards in the x direction by a scale factor $\frac{1}{k}$.

If $k < 1$ then the graph stretches outwards in the x direction by a scale factor $\frac{1}{k}$.

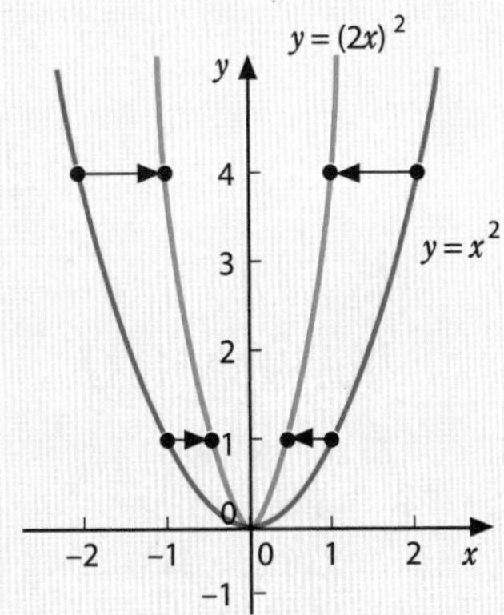

For this graph the y coordinates stay the same and the x values are multiplied by $\frac{1}{2}$.

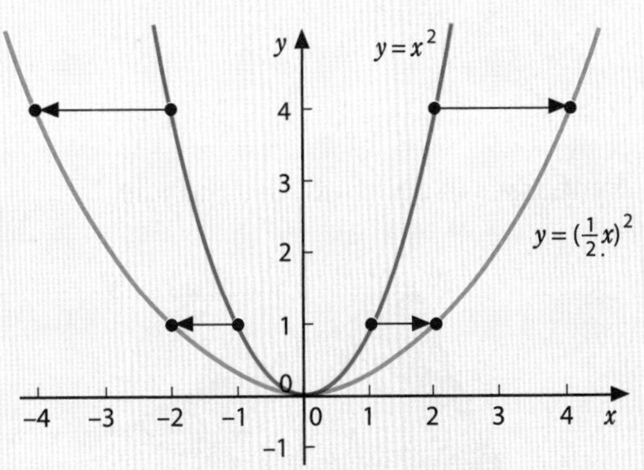

For this graph the y coordinates stay the same and the x values are multiplied by 2.

PROGRESS CHECK

1. The graph of $y = f(x)$ is shown on the grid.

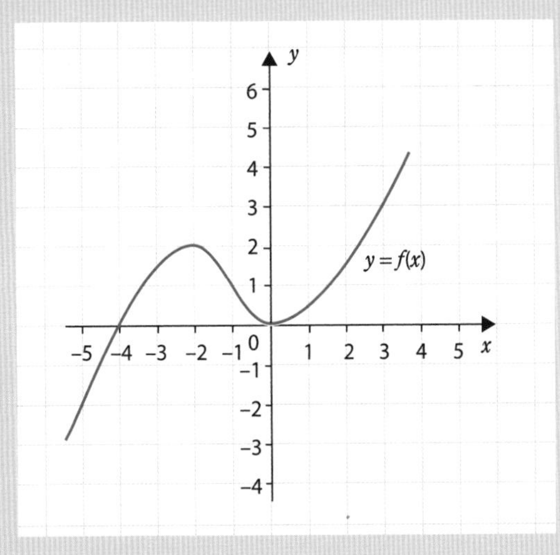

a. Sketch the graph of $y = f(x) + 1$

b. Sketch the graph of $y = -f(x)$

EXAM QUESTION

This is a sketch of the curve $y = f(x)$.

The minimum point of the curve is point B(2, –3).

Write down the coordinates of the minimum point for each of the following curves:

1. $y = f(x + 3)$
2. $y = f(x) - 4$
3. $y = f(x - 2)$
4. $y = f(-x)$
5. $y = f(x + 1)$

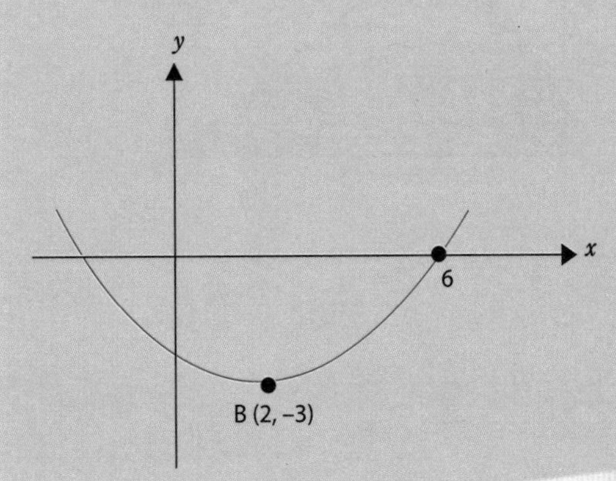

Constructions

The following constructions can be completed using only a ruler and a pair of compasses.

Constructing a triangle

Use compasses to construct this triangle.

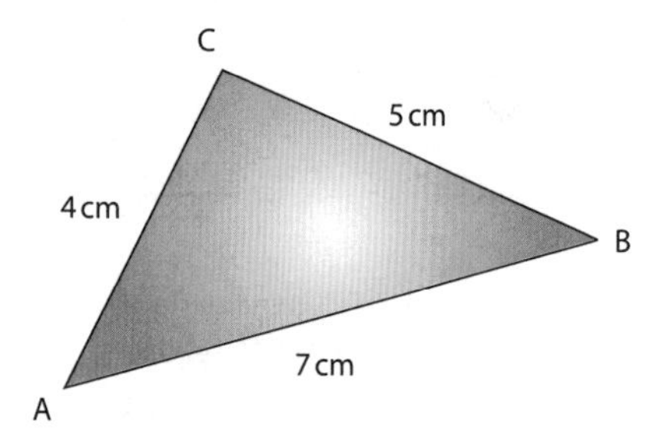

- Draw the longest side AB.
- With the compass point at A, draw an arc of radius 4 cm.
- With the compass point at B, draw an arc of radius 5 cm.
- Join A and B to the point where the two arcs meet at C.

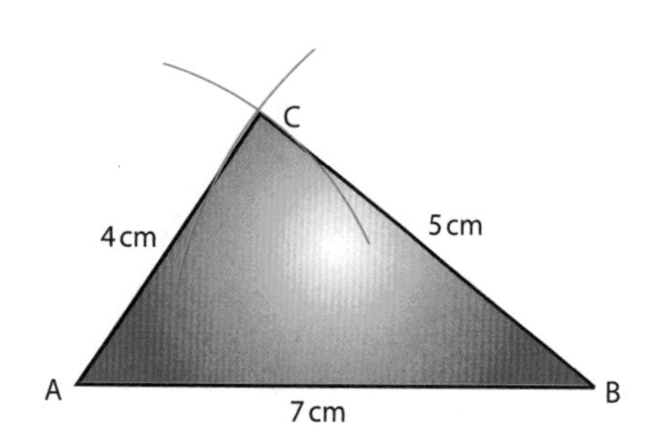

The perpendicular bisector of a line

- Draw a line XY.
- Draw two arcs with the compasses, using X as the centre. The compasses must be set at a radius greater than half the distance of XY.

- Draw two more arcs with Y as the centre.

  (Keep the compasses the same distance apart as before.)
- Join the two points where the arcs cross.
- AB is the **perpendicular bisector** of XY.
- N is the **midpoint** of XY.

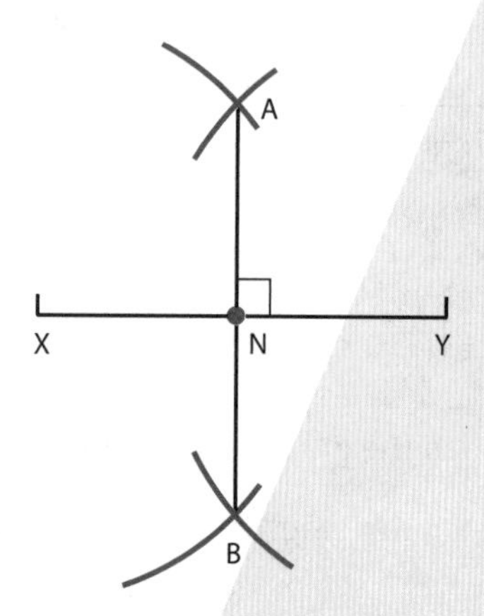

The perpendicular from a point to a line

- From P draw arcs to cut the line at A and B.
- From A and B draw arcs with the same radius to intersect at C.
- Join P to C; this line is perpendicular to AB.

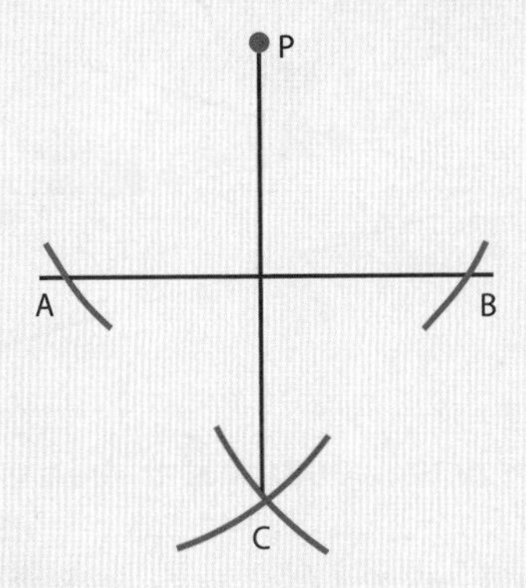

Bisecting an angle

- Draw two lines XY and YZ to meet at an angle.
- Using compasses, place the point at Y and draw arcs on XY and YZ.
- Place the compass point at the two arcs on XY and YZ and draw arcs to cross at N. Join Y and N. YN is the **bisector** of angle XYZ.

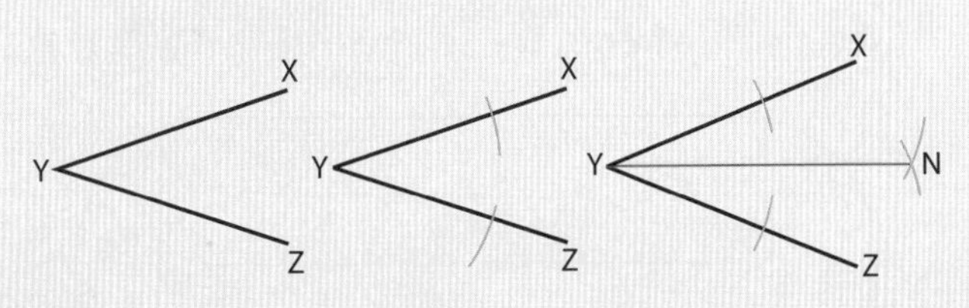

1. Construct an angle of 30°, using a ruler and pair of compasses only.
2. Draw the perpendicular bisector of an 8 cm line.
3. Bisect this angle.

40°

1. a. Using compasses only, construct the perpendicular bisector of the line AB. You must show all construction lines.

A B

b. Using compasses only, bisect this angle. You must show all construction lines.

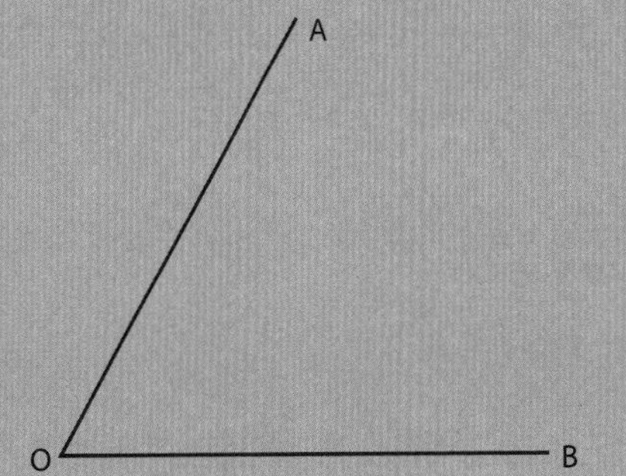

Loci

The locus of a point is the set of all the possible positions that the point can occupy subject to some given condition or rule.

Types of loci

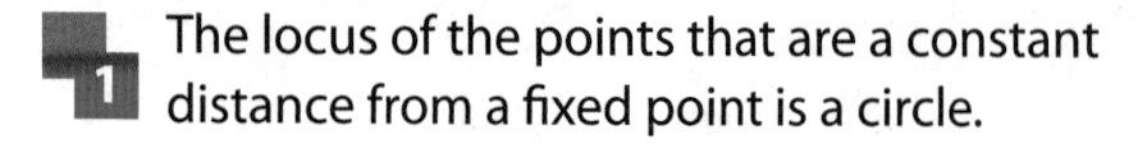

1 The locus of the points that are a constant distance from a fixed point is a circle.

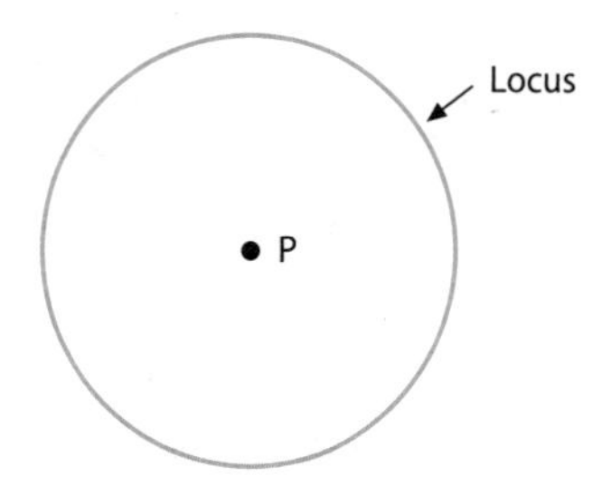

2 The locus of the points that are equidistant from two points XY is the perpendicular bisector of XY.

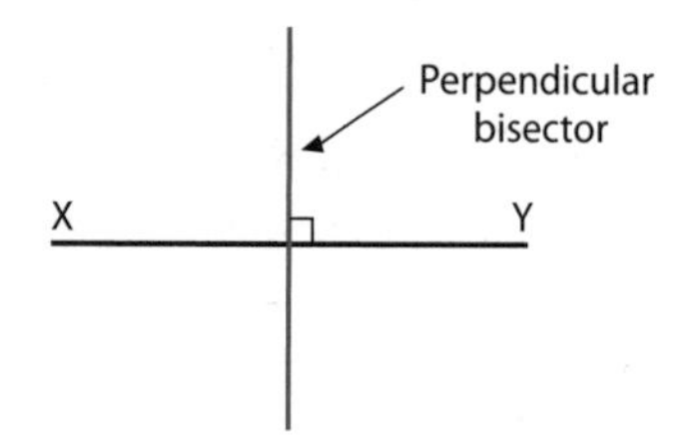

3 The locus of the points that are equidistant from two lines is the line that bisects the angle between the lines.

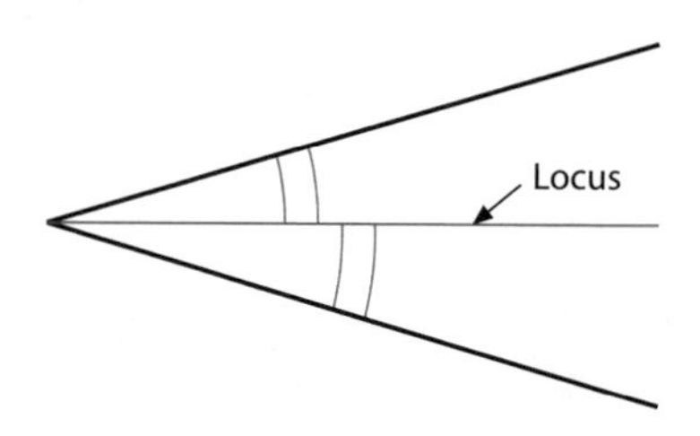

4 The locus of the points that are a constant distance from a line XY is a pair of parallel lines above and below XY.

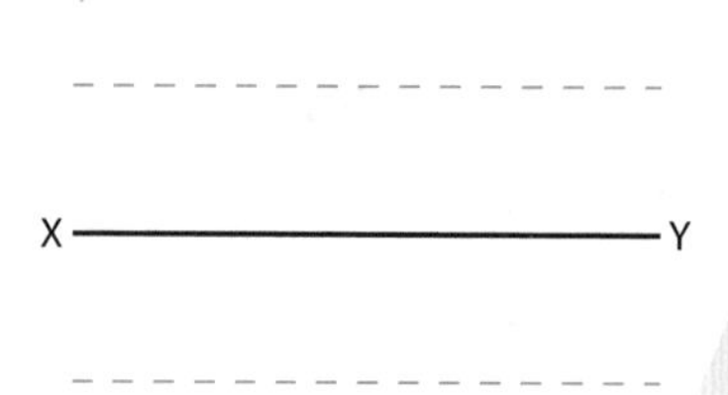

Sometimes you need to combine types 1 and 4.

A fixed distance from a line segment gives this locus.

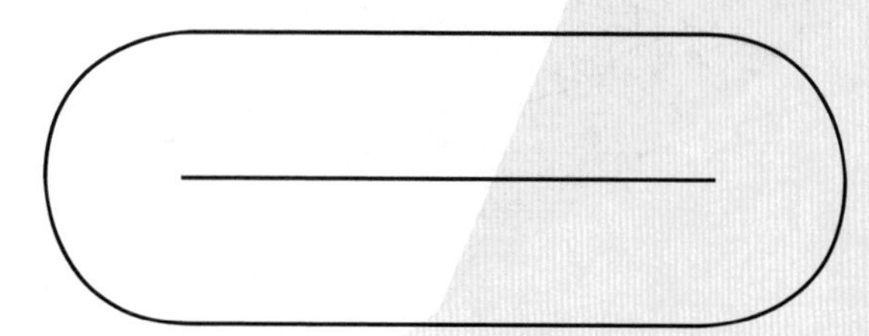

Example

Three radio transmitters form an equilateral triangle ABC with sides of 50 km. The range of the transmitter at A is 37.5 km, at B 30 km and at C 28 km. Using a scale of 1 cm to 10 km, construct a scale diagram to show where signals from all three transmitters can be received.

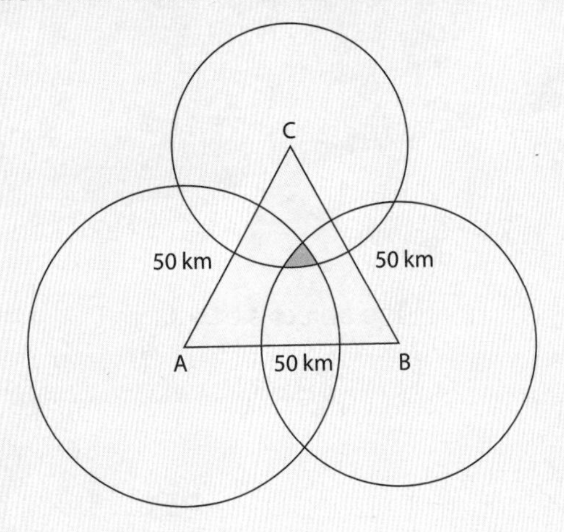

Please note that on your scale drawing the circle at A would have a radius of 3.75 cm. The circle at B would have a radius of 3 cm and the circle at C a radius of 2.8 cm.

PROGRESS CHECK

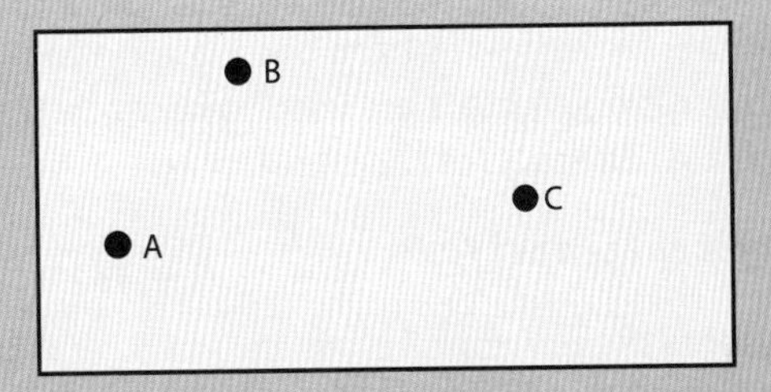

1. The plan shows a garden drawn to a scale of 1 cm : 2 m. A and B are bushes and C is a pond. A landscape-gardener has decided:

   a. to lay a path right across the garden at an equal distance from each of the bushes.

   b. to lay a flower border 4 m wide around pond C.

Construct these features on the plan above.

EXAM QUESTION

1. ABCD is a rectangle.

   Shade the set of points inside the rectangle which are both:

   more than 2 cm from point B and

   more than 1.5 cm from the line AD.

Translations and Reflections

There are four types of transformations: translation, reflection, rotation and enlargement.

Translations

Translations move figures from one position to another position. **Vectors** are used to describe the distance and direction of the translations.

A vector is written as $\begin{pmatrix} a \\ b \end{pmatrix}$

a represents the horizontal distance and ***b*** represents the vertical distance

The object and the image are **congruent** when the shape is translated.

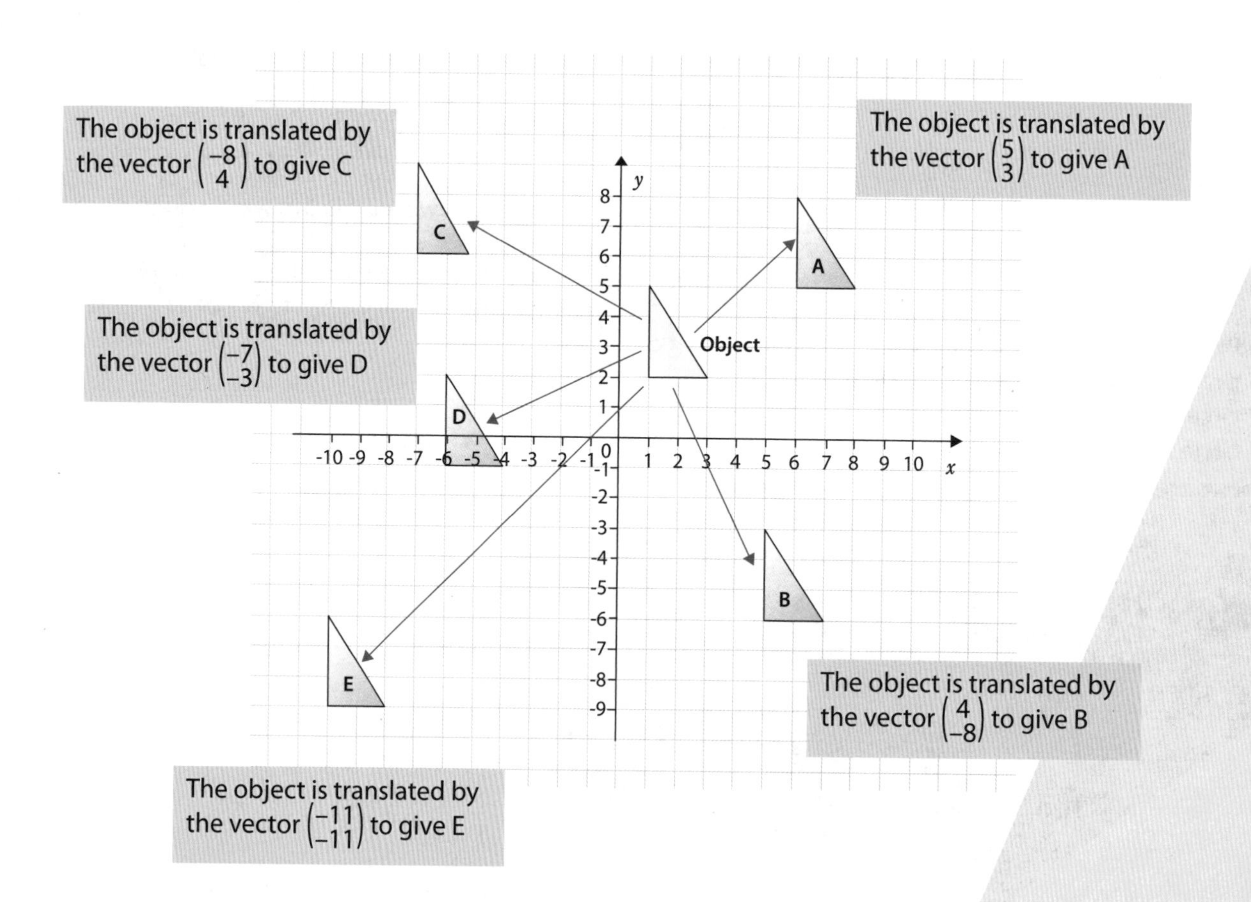

Reflections

These create an image of an object on the other side of the mirror line.

The mirror line is known as an **axis of reflection.**

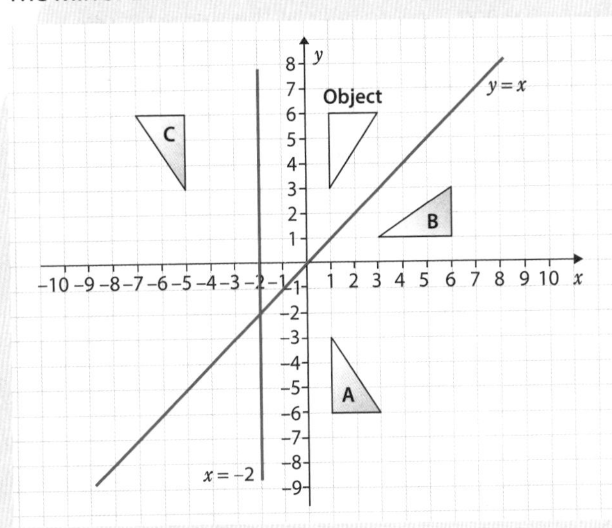

The object is reflected in the x-axis (or $y = 0$) to give the image A.

The object is reflected in the line $y = x$ to give the image B.

The object is reflected in the line $x = -2$ to give the image C.

PROGRESS CHECK

For the diagram below, describe fully the transformation that maps:

1. A onto B
2. B onto C
3. A onto D
4. A onto E

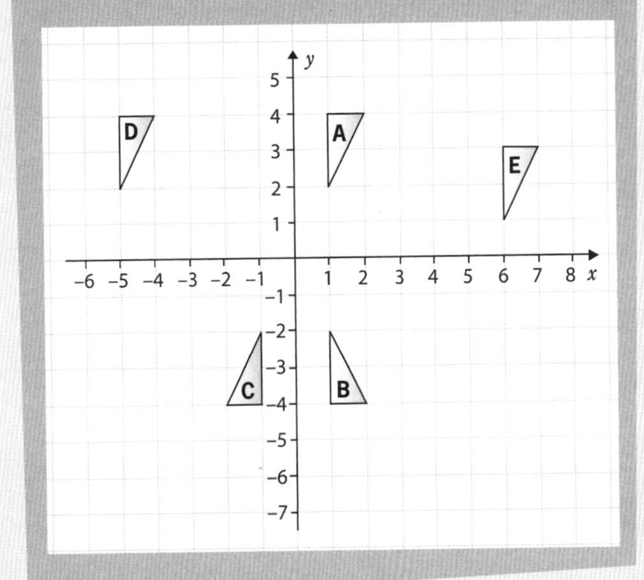

EXAM QUESTION

1. Triangle A and triangle B have been drawn on the grid.
   a. Reflect triangle A in the line $x = 5$
      Label this image C.
   b. Translate triangle B by the vector $\begin{pmatrix} 4 \\ 2 \end{pmatrix}$
      Label this image D.
   c. Describe fully the single transformation which will map triangle A onto triangle B.

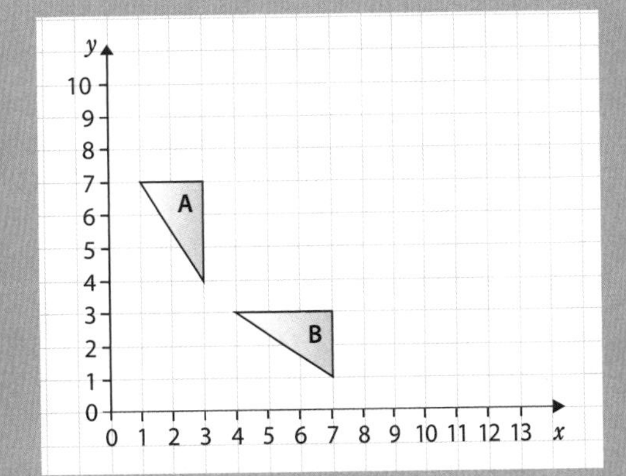

Rotation and Enlargement

After a rotation, the image is congruent to the object. After an enlargement, the enlarged shape is similar to the object.

Rotations

In a rotation the object is turned by a given angle about a fixed point called the **centre of rotation**. The size and shape of the figure are not changed.

The object A is rotated by 90° clockwise about (0, 0) to give the image B.

The object A is rotated by 180° about (0, 0) to give the image C.

The object A is rotated 90° anticlockwise about (–2, 2) to give the image D.

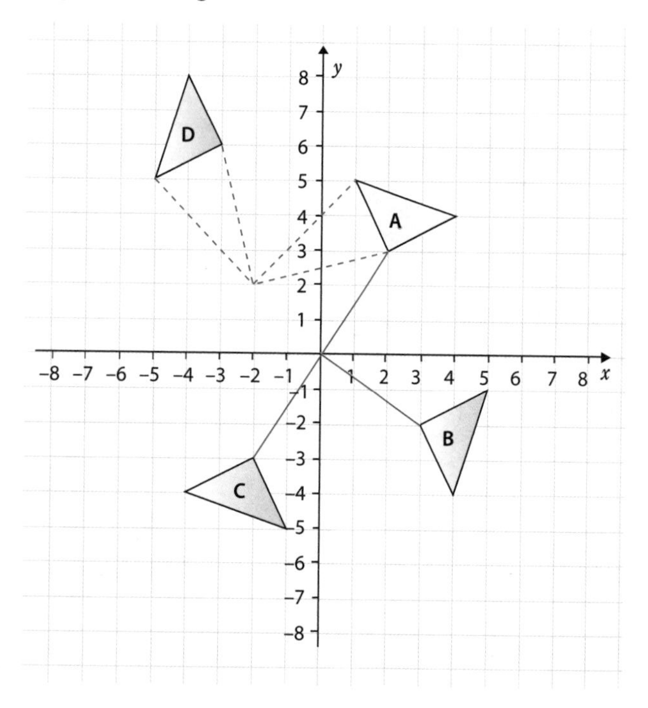

Enlargements

These change the size but not the shape of the object. The **centre of enlargement** is the point from which the enlargement takes place. The **scale factor** indicates how many times the lengths of the original figure have changed size.

Example

Describe fully the transformation that maps ABC onto A′B′C′.

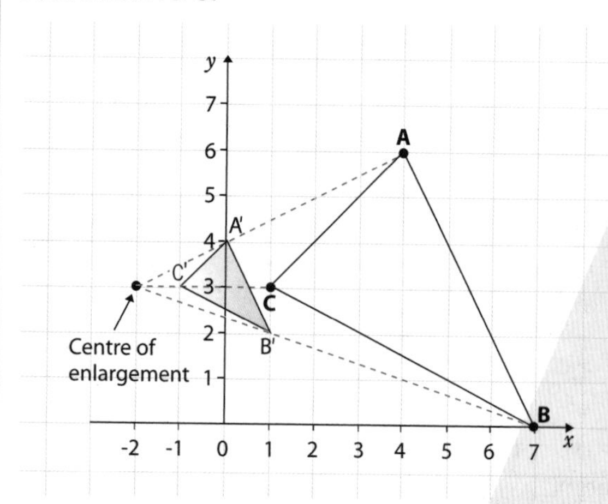

- To find the centre of enlargement join A to A′, B to B′ etc. and continue the line.
- Where all the lines meet is the centre of enlargement: (–2, 3)
- The transformation is an enlargement of scale factor $\frac{1}{3}$. Centre of enlargement is (–2, 3).

An enlargement with a scale factor less than 1 makes the shape smaller.

Enlargements with a negative scale factor

For an enlargement with a negative scale factor, the image is situated on the opposite side of the centre of enlargement.

The triangle A (3, 4), B (9, 4) and C (3, 10) is enlarged with a scale factor of $\frac{-1}{3}$, about the centre of enlargement (0, 1). Label the enlargement A'B'C'.

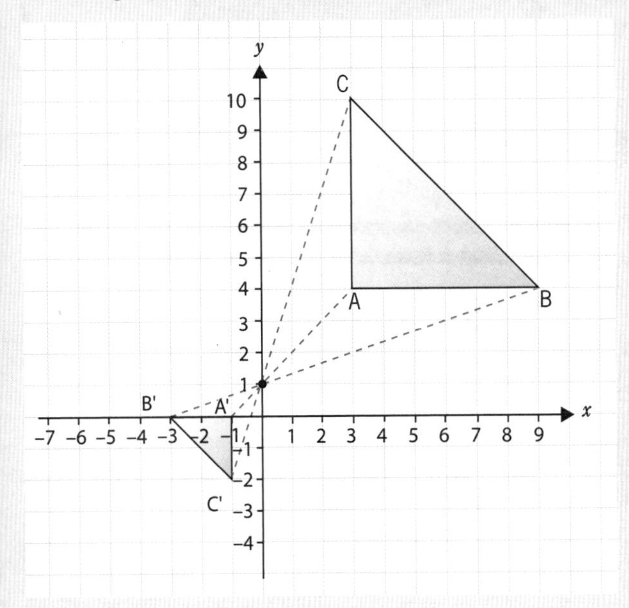

- Notice A'B'C' is a third of the size of triangle ABC.
- ABC is on the opposite side of the centre of enlargement.

1. Complete the diagram below, to show the enlargement of the shape by a scale factor of $\frac{1}{2}$. Centre of enlargement at (0, 0). Call the shape T.

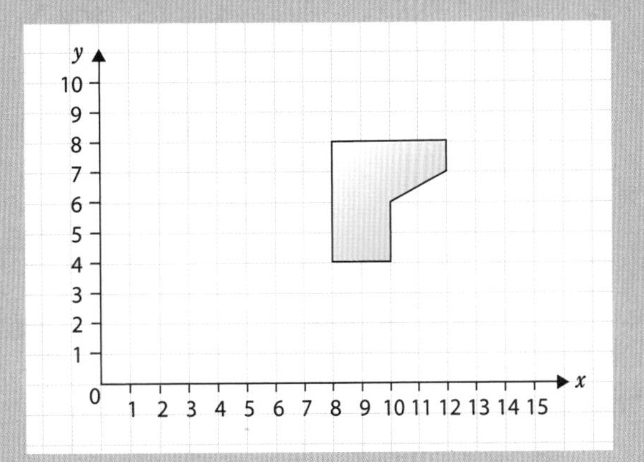

EXAM QUESTION

1. The quadrilateral is enlarged with a scale factor of $\frac{-1}{2}$ about the origin. Draw the enlargement on the diagram.

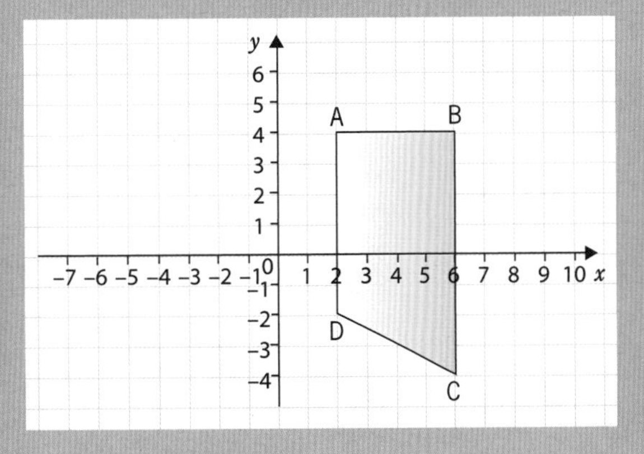

Similarity

Objects that are exactly the same shape but different sizes are called similar shapes. One is an enlargement of the other.

Similarity

Corresponding angles are equal.

Corresponding lengths are in the same ratio.

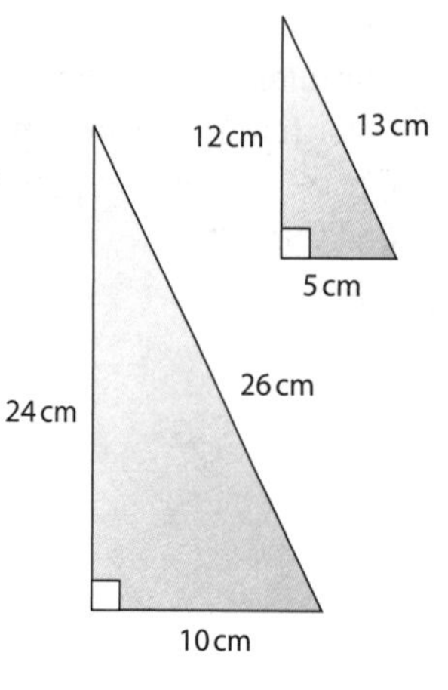

Finding missing lengths of similar figures

These questions are very common at GCSE.

Example

Find the missing lengths labelled a in the diagrams below:

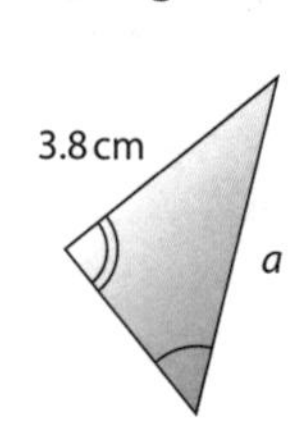

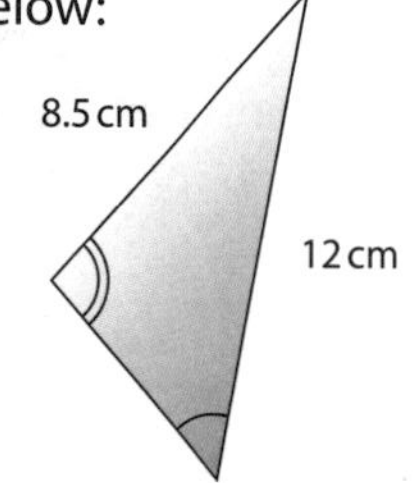

a. $\frac{a}{12} = \frac{3.8}{8.5}$

- Corresponding lengths are in the same ratio.

$a = \frac{3.8}{8.5} \times 12$

$a = 5.36\text{ cm (3sf)}$

Multiply both sides by 12.

Example

b. $\frac{a}{7.2} = \frac{19.5}{13.1}$

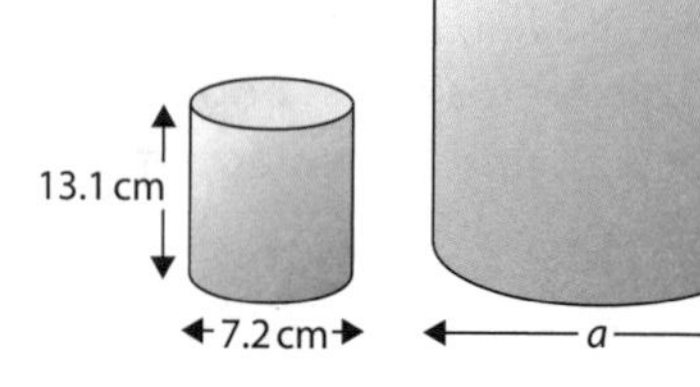

$a = \frac{19.5}{13.1} \times 7.2$

$a = 10.7\text{cm (3sf)}$

Multiply both sides by 7.2.

Example

Calculate the missing length y.

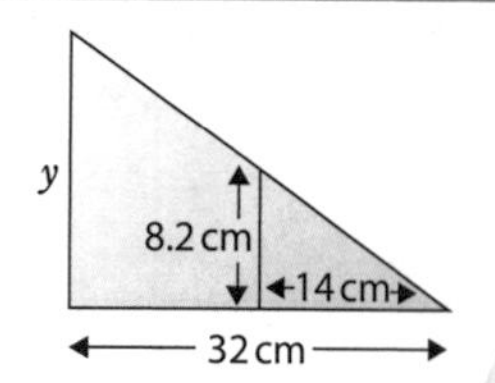

- Firstly draw out the individual triangles:

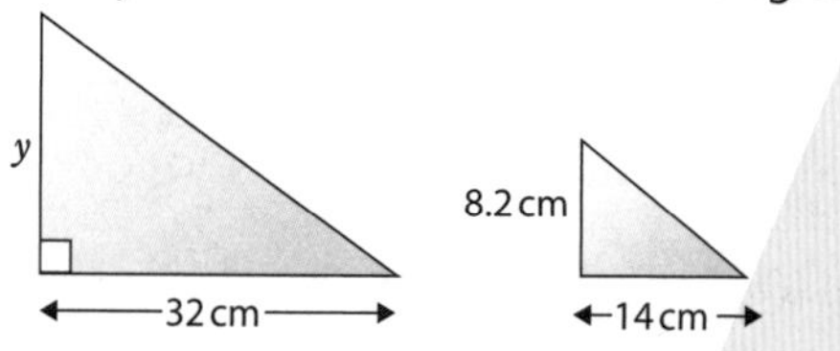

- Write down the corresponding ratios:

$\frac{y}{32} = \frac{8.2}{14}$

- Multiply both sides by 32:

$y = \frac{8.2}{14} \times 32$

$= 18.7\text{cm}$

This gives an alternative way to writing the ratios as seen in the other examples. Both are correct!

Area and volume of similar figures

Areas

Areas of similar figures are not in the same ratio as their lengths.

If the corresponding lengths are in the ratio $a : b$ then their areas are in the ratio $\boldsymbol{a^2 : b^2}$.

Volumes

A similar result can be found when looking at the volumes of similar figures.

If the corresponding lengths are in the ratio $a : b$, then their volumes are in the ratio $\boldsymbol{a^3 : b^3}$.

For a scale factor n:
The sides are n times bigger
The areas are n^2 times bigger
The volumes are n^3 times bigger

Example
These two solids are similar. If the volume of the smaller solid is 9 cm³, calculate the volume of the larger solid.

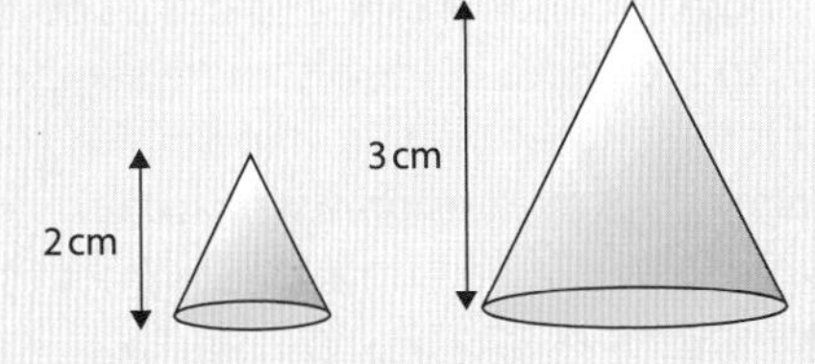

Linear scale factor is 2 : 3

Volume scale factor is $2^3 : 3^3$

Volume of larger solid is $\frac{27}{8} \times 9$

$= 30.375\ \text{cm}^3$

PROGRESS CHECK

1. Calculate the lengths marked n in these similar shapes. Give your answers correct to 1dp.

a.

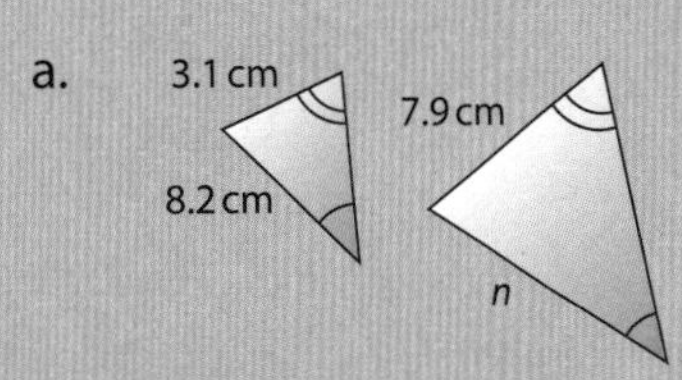

b.

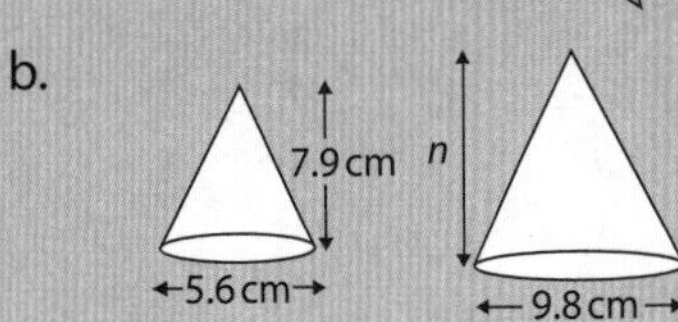

c.

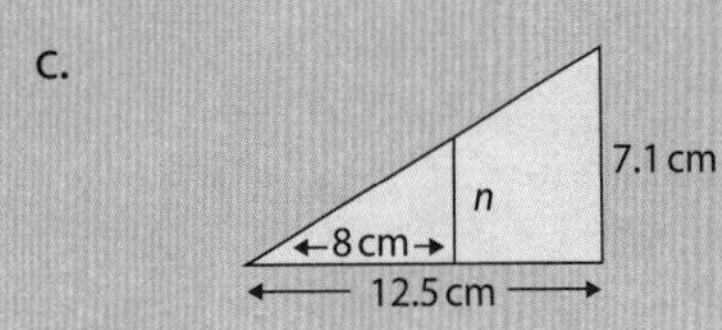

EXAM QUESTION

1. The heights of two similar shapes are 8 cm and 12 cm. If the area of the larger shape is 64 cm², find the area of the smaller shape.

Circle theorems

There are several theorems about circles that you need to know.

The circle theorems

1 The perpendicular bisector of any chord passes through the centre of the circle.

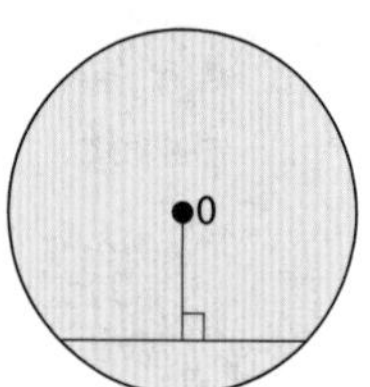

2 The angle in a semicircle is always 90°.

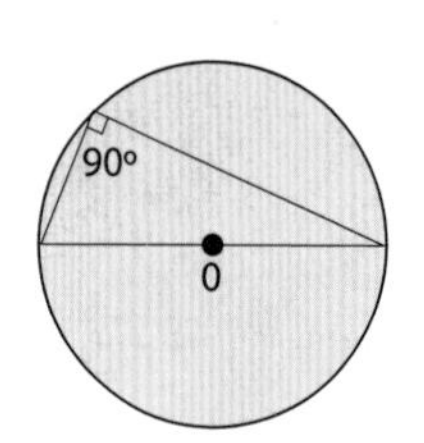

3 The radius and a tangent always meet at 90°.

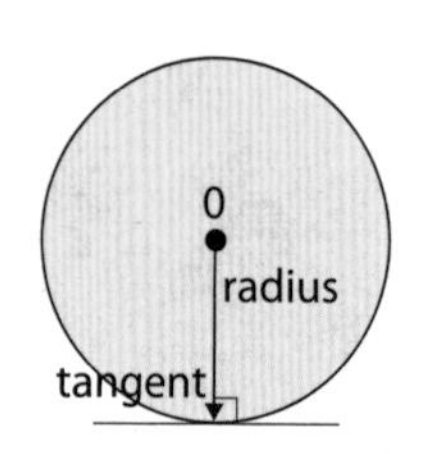

4 Angles in the same segment are equal, e.g. $A\hat{B}C = A\hat{D}C$

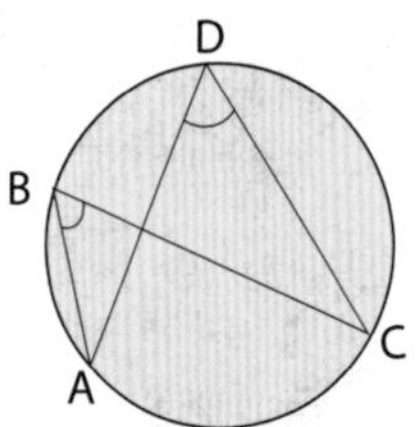

5 The angle at the centre is twice the angle at the circumference, e.g. $P\hat{O}Q = 2 \times P\hat{R}Q$

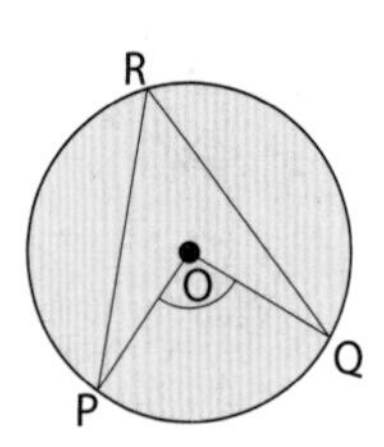

6 Opposite angles of a cyclic quadrilateral add up to 180°.

(A cyclic quadrilateral is a 4-sided shape with each vertex touching the circumference of the circle.)

i.e. $x + y = 180°$

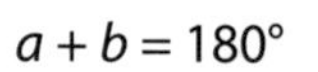

$a + b = 180°$

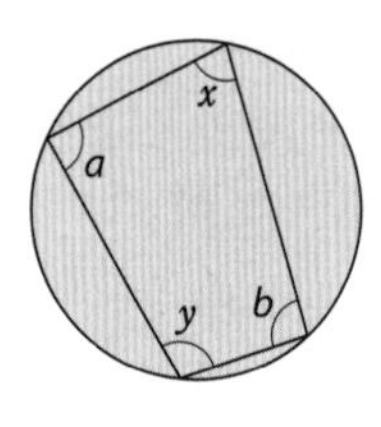

7 The lengths of two tangents from a point are equal, e.g. RS = RT

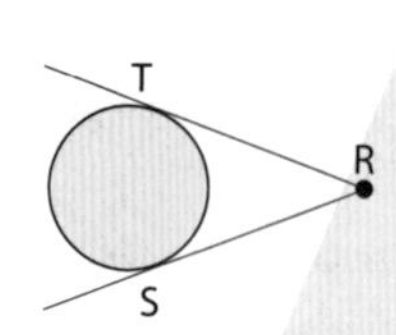

8 The angle between a tangent and a chord is equal to the angle in the alternate segment (that is the angle which is made at the edge of the circle by two lines drawn from the chord). This is known as the **Alternate Segment Theorem.** $a = b$

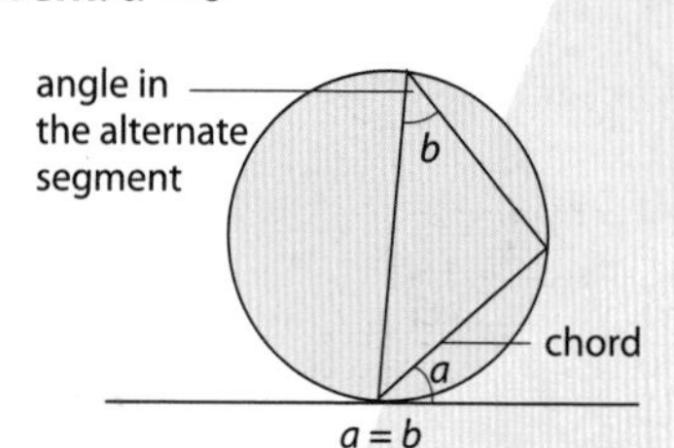

Example

Calculate the angles marked a–d in the diagram below. Give a reason for your answers.

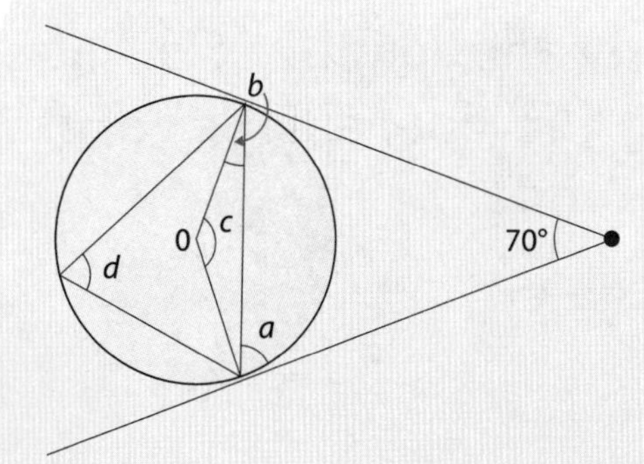

$a = \dfrac{180 - 70°}{2}$	**Angles in a triangle add up to 180°.**
$= \dfrac{110°}{2}$	**Isosceles triangle, base angles are equal.**
$= 55°$	
$b = 90° - 55°$	**Radius and tangent meet at 90°.**
$= 35°$	
$c = 180° - (2 \times 35°)$	**Angles in a triangle add up to 180°.**
$= 110°$	
$d = 110° \div 2$	**Angle at the centre is twice the angle at the circumference.**
$= 55°$	

PROGRESS CHECK

Some angles are written on cards. Match the missing angles in the diagrams below with the correct card. 0 represents the centre of the circle.

53° 50° 62° 109° 126°

a.

b.

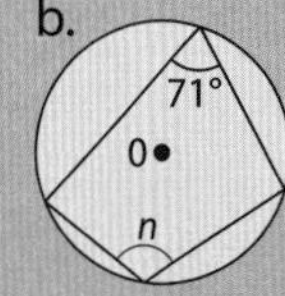

c.

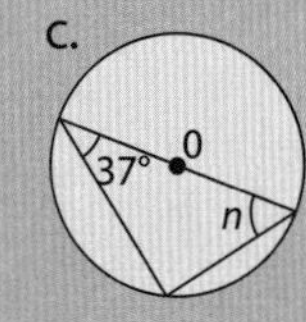

d.

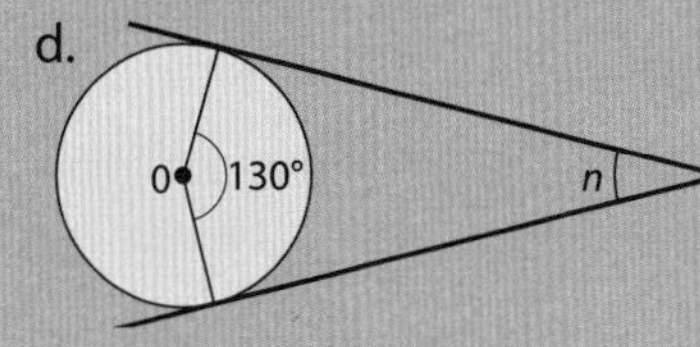

e.

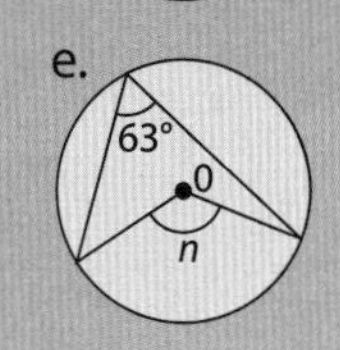

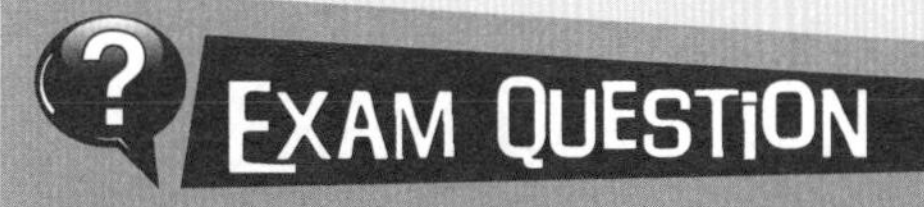

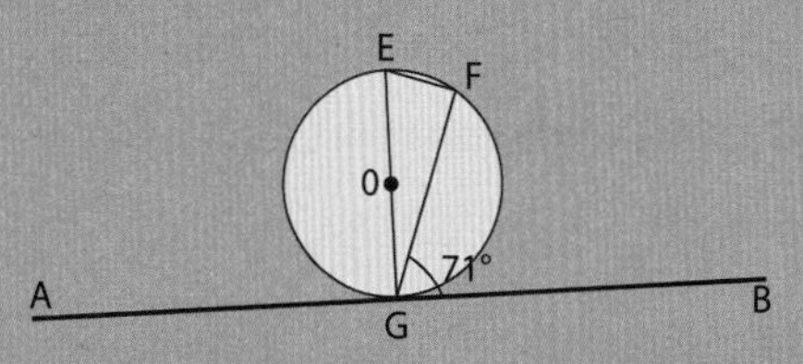

1. In the diagram, E, F and G are points on the circle, centre 0.

   $F\hat{G}B = 71°$

   AB is a tangent to the circle at point G.

   a. Calculate the size of angle $E\hat{G}F$.
      Give reasons for your answer.

   b. Calculate the size of angle $G\hat{E}F$.
      Give reasons for your answer.

Pythagoras' Theorem

Pythagoras' Theorem states that 'The square on the hypotenuse is equal to the sum of the squares on the other two sides.'

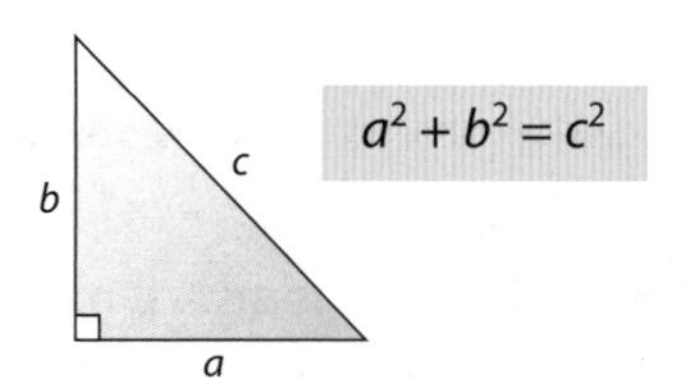

Finding the hypotenuse

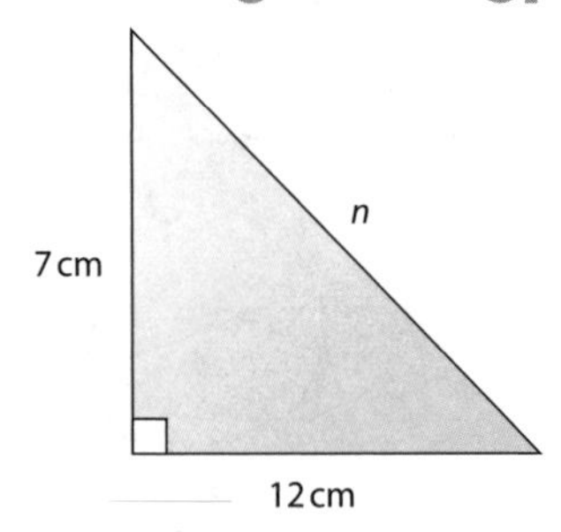

$n^2 = 7^2 + 12^2$	Square the two sides.
$n^2 = 49 + 144$	Add the two sides together.
$n = 193$	
$n = \sqrt{193}$	Square root.
$n = 13.9$ cm (3sf)	Round to 3sf.

Finding a short side

$15^2 = p^2 + 8^2$

$15^2 - 8^2 = p^2$

$225 - 64 = p^2$

$161 = p^2$

$\sqrt{161} = p$

$p = 12.7$ cm (3sf)

15 cm
p
8 cm

When finding a shorter length remember to subtract.

Finding the length of a line AB, given the coordinates of its end points

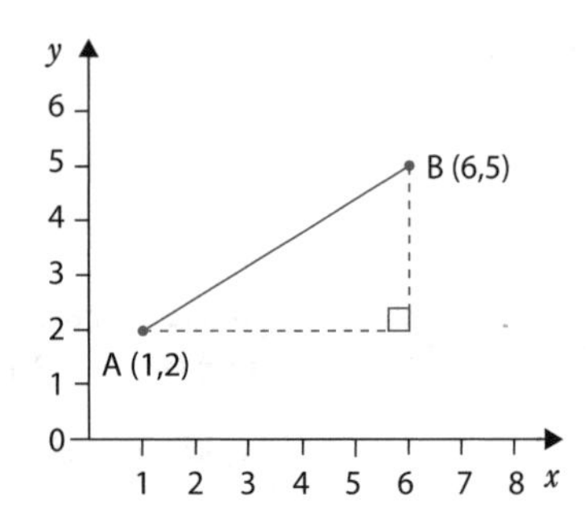

Horizontal distance $= 6 - 1 = 5$

Vertical distance $= 5 - 2 = 3$

Length of $(AB)^2 = 5^2 + 3^2$

$= 25 + 9$

$= 34$

$AB = \sqrt{34}$

$AB = 5.83$ cm

We could leave this as $\sqrt{34}$. This is known as leaving in **surd form**.

Solving a more difficult problem

Calculate the vertical height of this isosceles triangle.

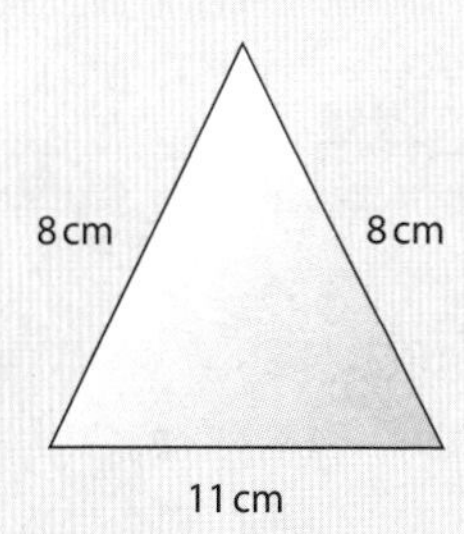

Remember to split the triangle down the middle to make it right-angled.

Using Pythagoras' Theorem gives:

$$8^2 = h^2 + 5.5^2$$

$$64 = h^2 + 30.25$$

$$64 - 30.25 = h^2$$

$$33.75 = h^2$$

$$\sqrt{33.75} = h^2$$

$$h = 5.81 \text{ cm (3sf)}$$

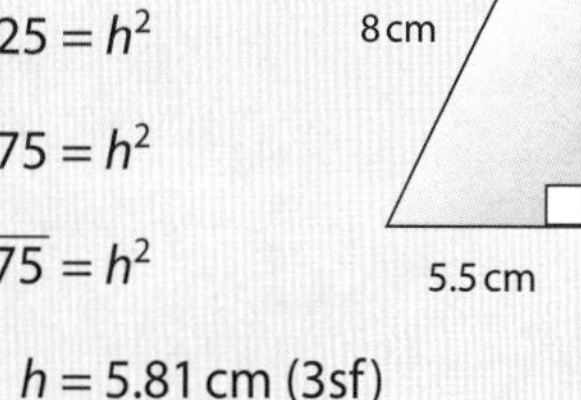

Progress Check

1. Molly says: 'The angle x in this triangle is 90°.'

Explain how Molly knows that without measuring the size of the angle.

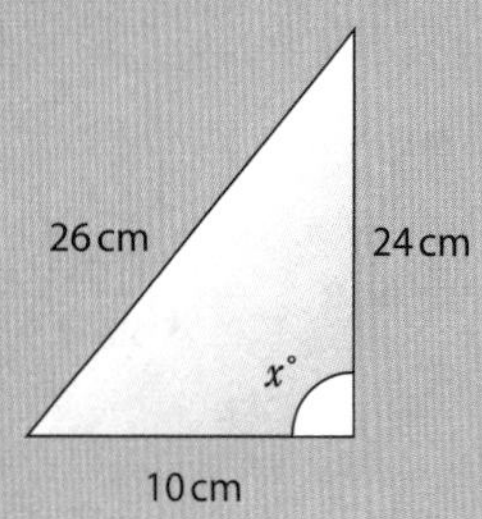

Exam Question

1. Calculate the length of the diagonal of this rectangle.

Give your answer to one decimal place (1dp).

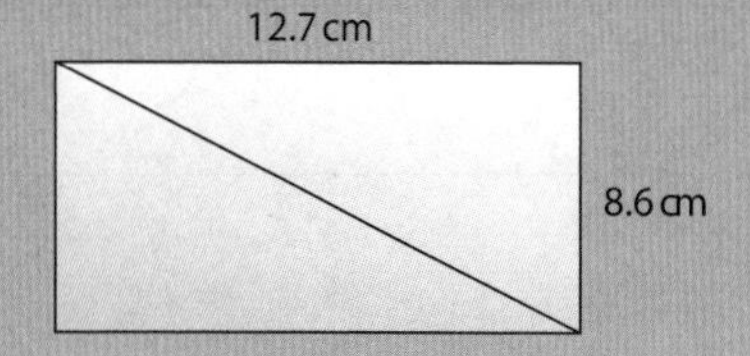

2. Calculate the length of CD in this diagram.

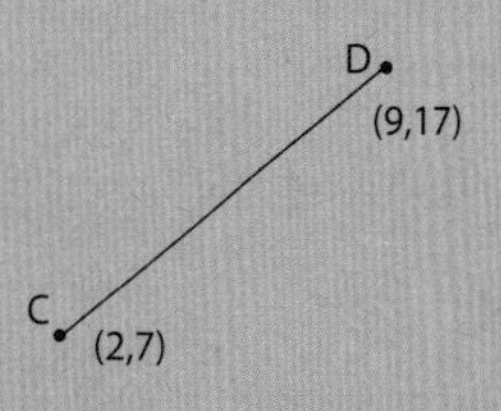

Leave your answer in surd form.

Trigonometry

Trigonometry in right-angled triangles can be used to find an unknown angle or length.

The sides of a right-angled triangle are given temporary names according to where they are in relation to a chosen angle θ.

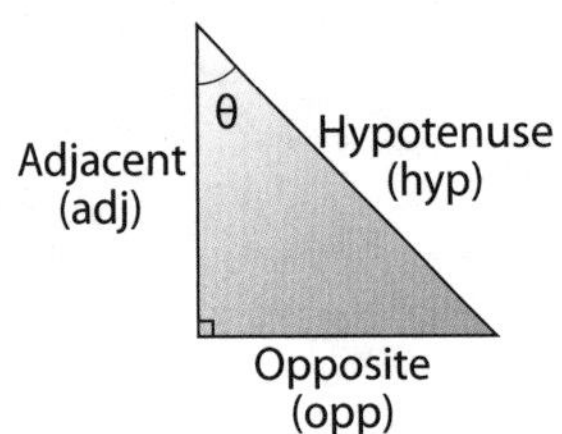

The trigonometric ratios are:

$$\text{Sine } \theta = \frac{\text{Opposite}}{\text{Hypotenuse}} \quad \text{Cosine } \theta = \frac{\text{Adjacent}}{\text{Hypotenuse}}$$

$$\text{Tangent } \theta = \frac{\text{Opposite}}{\text{Adjacent}}$$

Use the words **SOH – CAH – TOA** to remember the ratios.

Example: TOA means $\tan\theta = \frac{\text{opp}}{\text{adj}}$

Finding a length

Example

Find the missing length y in the diagram.

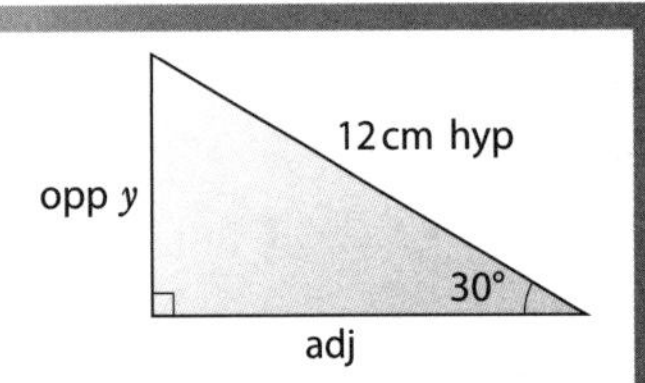

- Label the sides first.
- Decide on the ratio.

  $\sin 30° = \frac{\text{opp}}{\text{hyp}}$

- Substitute in the values

  $\sin 30° = \frac{y}{12}$

  $12 \times \sin 30° = y$

  $y = 6\,\text{cm}$

Multiply both sides by 12.

Finding an angle

Example

Calculate angle ABC.

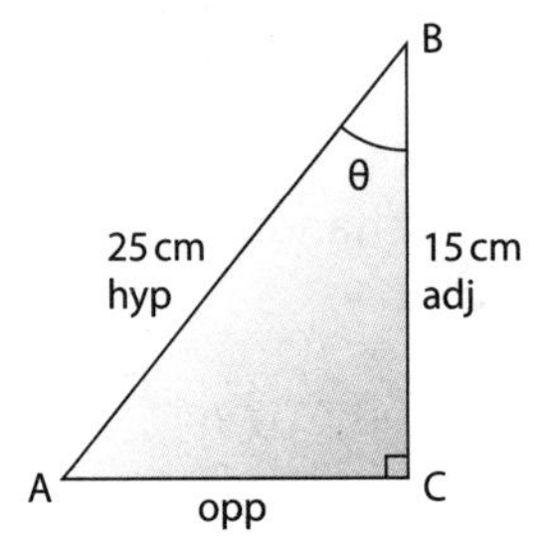

Label the sides and decide on the ratio.

$\cos\theta = \frac{\text{adj}}{\text{hyp}}$

$\cos\theta = \frac{15}{25}$

$\cos\theta = 0.6$

$\theta = \cos^{-1} 0.6$

$= 53.13°$

To find the angle, you usually use the second function on your calculator.

It is important you know how to use your calculator when working out trigonometry questions.

Angle of Elevation

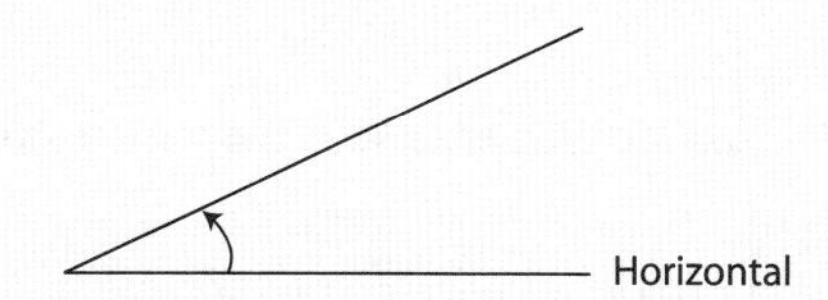

This is the angle measured from the horizontal upwards.

Angle of Depression

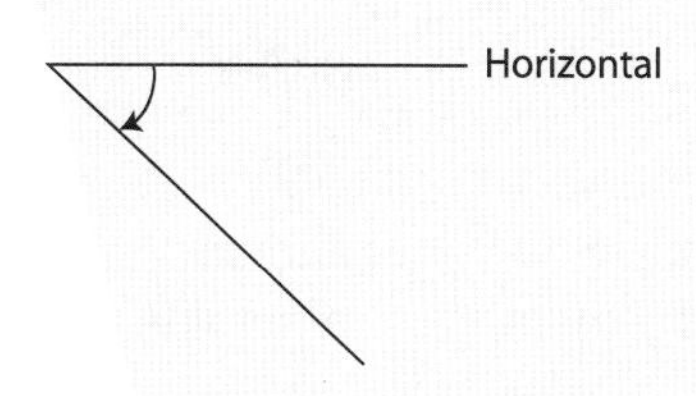

This is the angle measured from the horizontal downwards.

PROGRESS CHECK

1. Work out the missing lengths labelled x in the diagrams below:

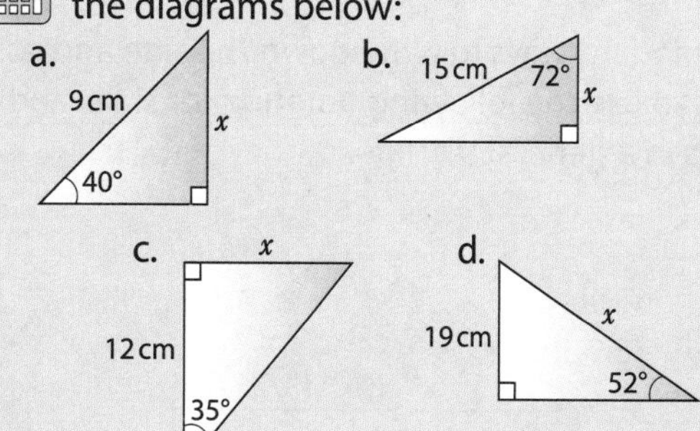

2. Work out the missing angles labelled x in the diagrams below:

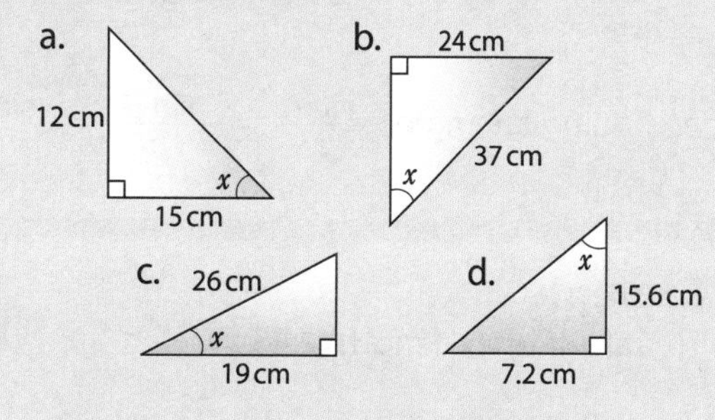

EXAM QUESTION

1. The diagram represents a vertical mast, PN. The mast is supported by two metal cables, PA and PB, fixed to the horizontal ground at A and B.

BN = 12.6 m

PN = 19.7 m

angle PAN = 48°

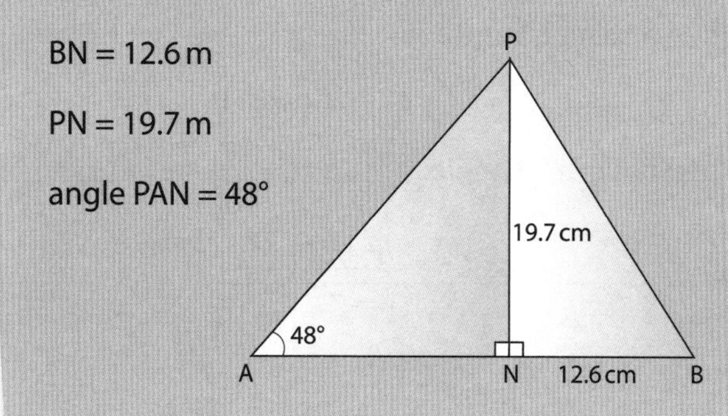

a. Calculate the size of angle PBN.

Give your answer correct to three significant figures.

b. Calculate the length of the metal cable PA.

Give your answer correct to 3 significant figures.

Sine and Cosine rule

The sine and cosine rules allow you to solve problems in triangles that do not contain a right angle.

The sine rule

The standard ways to write down the sine and cosine rule is to use the following notation for sides and angles in a general triangle.

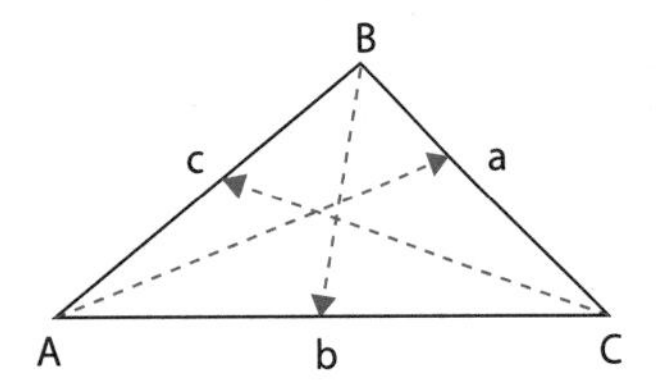

When finding a **length**: $\dfrac{a}{\sin A} = \dfrac{b}{\sin B} = \dfrac{c}{\sin C}$

When finding an **angle**: $\dfrac{\sin A}{a} = \dfrac{\sin B}{b} = \dfrac{\sin C}{c}$

Example

1. Calculate the length of RS.

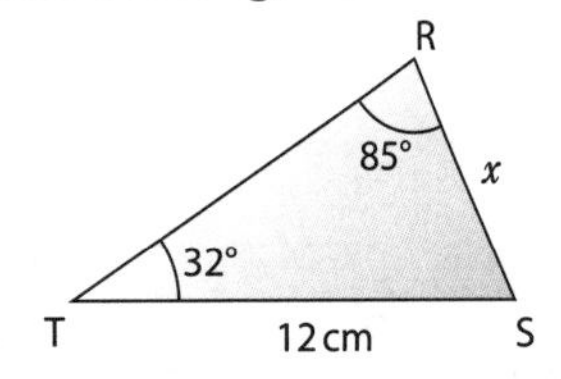

- Call RS, x, the length to be found.
- Since we have two angles and a side we use the sine rule

$$\frac{x}{\sin 32°} = \frac{12}{\sin 85°}$$

- Rearrange to make x the subject:

$$x = \frac{12}{\sin 85°} \times \sin 32°$$

Length of RS = 6.38 cm (3sf)

Example

2. Calculate the size of $C\hat{D}E$.

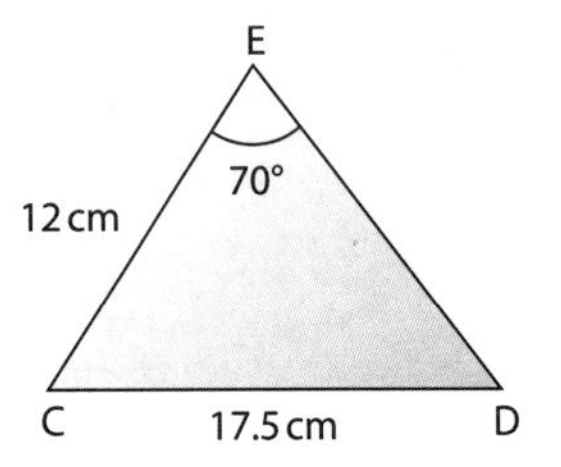

- Since we have two sides and an angle not enclosed by them, we use the sine rule.

$$\frac{\sin D}{12} = \frac{\sin 70°}{17.5}$$

- Rearrange to give:

$$\sin D = \frac{\sin 70°}{17.5} \times 12$$

$$\sin D = 0.644\ldots$$

$$D = \sin^{-1} 0.644$$

$$D = 40.1°$$

The cosine rule

When finding a **length**:

$$a^2 = b^2 + c^2 - (2bc \cos A)$$

When finding an **angle**:

$$\cos A = \frac{b^2 + c^2 - a^2}{2bc}$$

Example

Calculate the length of JK.

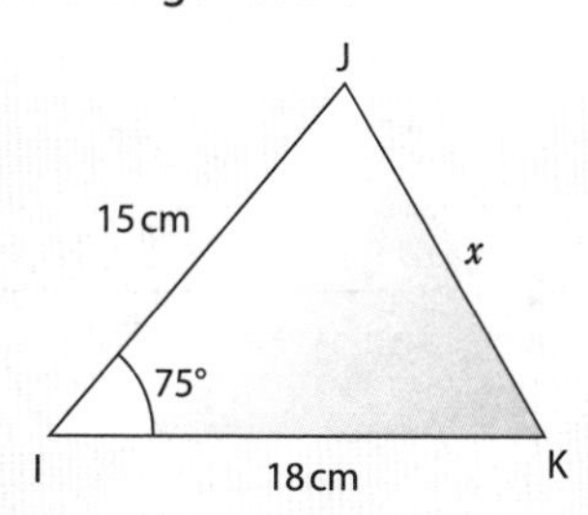

- Call JK, x, the length to be found.
- Since we have two sides and the included angle, we use the cosine rule.

$x^2 = 15^2 + 18^2 - (2 \times 15 \times 18 \times \cos 75°)$

$x^2 = 409.23\ldots$

$x = \sqrt{409.23}$

JK = 20.2 cm (3sf)

Progress Check

1. For the questions below decide whether the missing length x is correct.

a. $x = 11.49$ cm

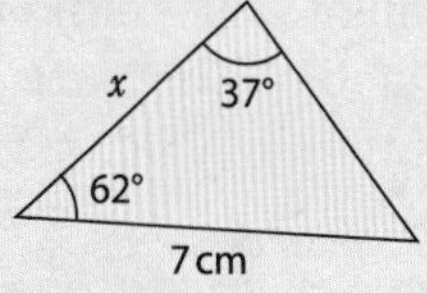

b. $x = 20.93$ cm

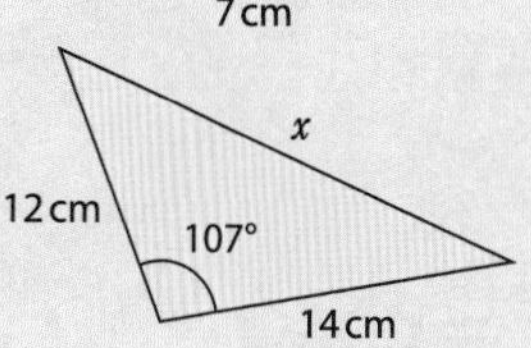

c. $x = 13.87$ cm

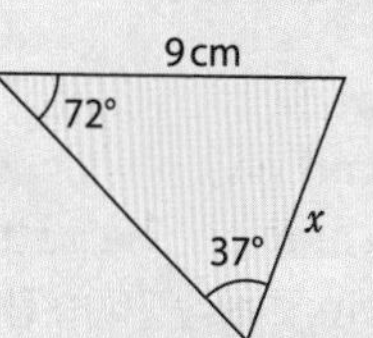

2. Calculate the missing angles in the diagrams below:

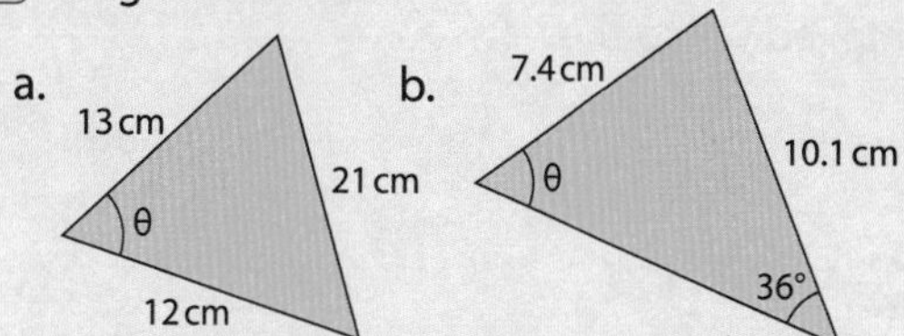

Exam Question

1. The diagram shows a vertical pole AB on horizontal ground BCD.

BCD is a straight line.

The angle of elevation of A from C is 58°.
The angle of elevation of A from D is 32°.

Calculate the height of the pole.

Give your answer correct to 3 significant figures.

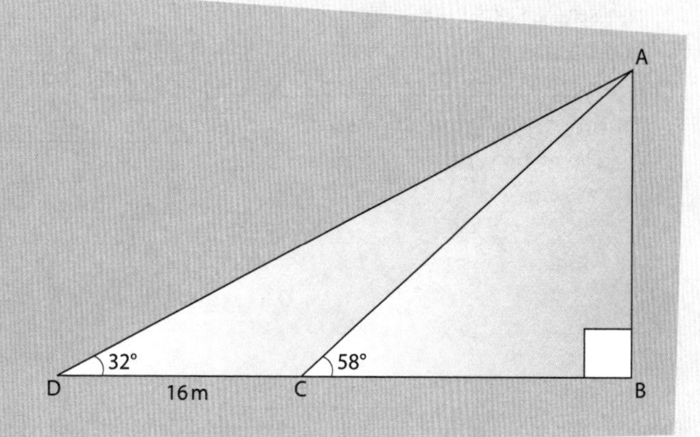

Trigonometric functions

The behaviour of the sine, cosine and tangent functions may be represented graphically.

$y = \sin x$

The maximum and minimum values of $\sin x$ are 1 and –1. The pattern repeats every 360°.

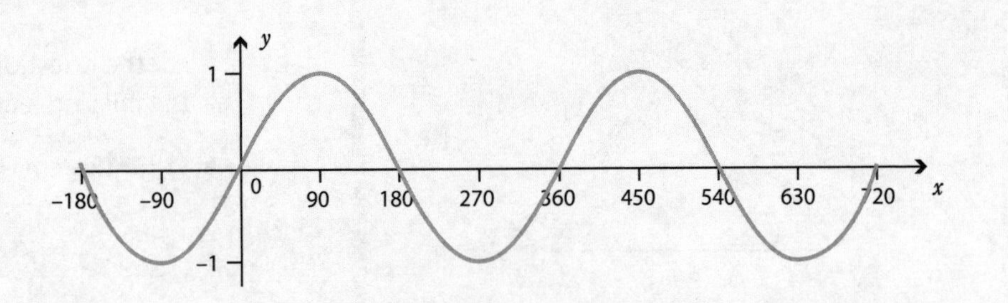

$y = \cos x$

The maximum and minimum values of $\cos x$ are 1 and –1. The pattern repeats every 360°. This graph is the same as $y = \sin x$ except it has been moved 90° to the left.

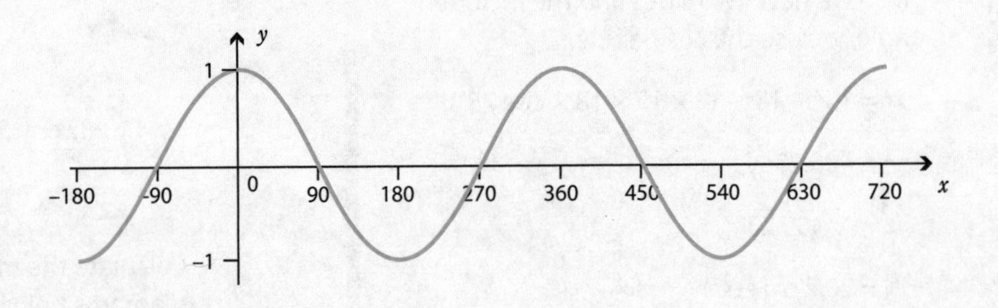

$y = \tan x$

This graph is nothing like the sine or cosine curve. The values of $\tan x$ repeat every 180°. The tan of 90° is infinity, i.e. a value so great it cannot be written down.

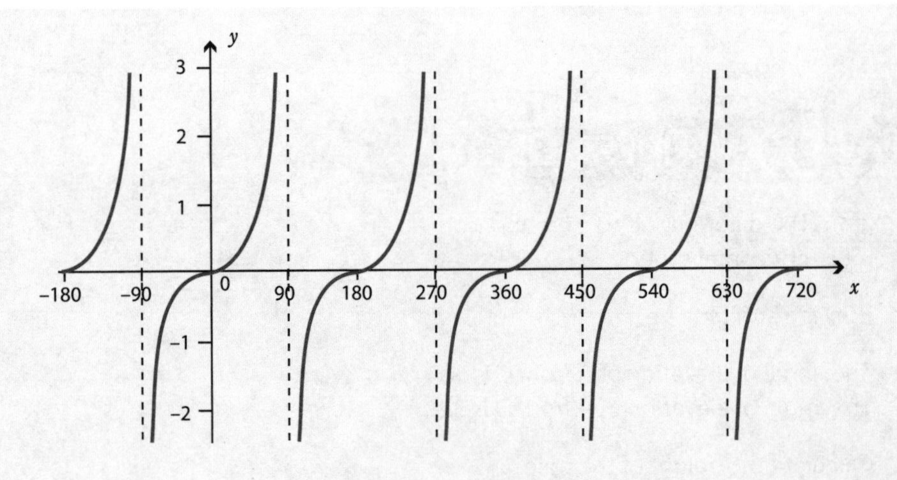

The trigonometric graphs can be used to solve inverse problems.

Example

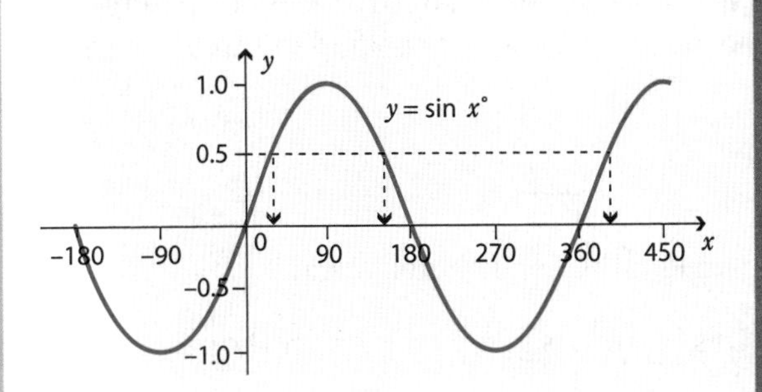

To solve $\sin x = 0.5$ for values of x between 0 and 360°, draw a straight line across at $y = 0.5$ and by its symmetrical properties we can see that the solutions are $x = 30°, 150°$, (390° is out of the range).

PROGRESS CHECK

1. On the axes below draw the graph of $y = \sin 2x$.

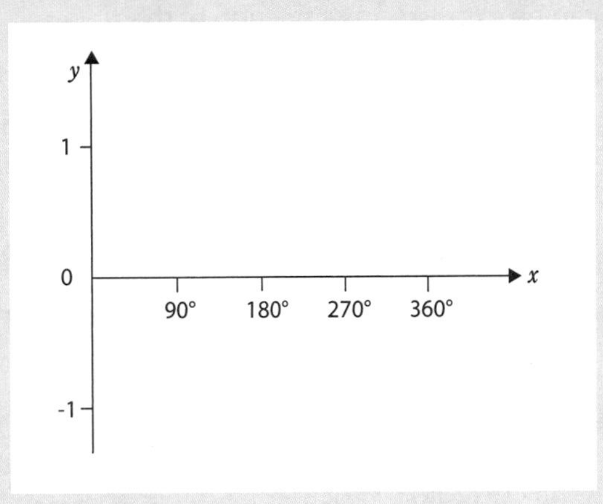

2. If $\cos \theta = 0.5$, write down the values of θ between 0° and 360°.

EXAM QUESTION

1. 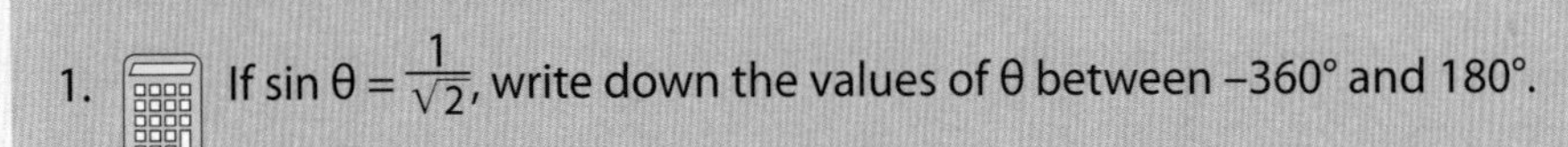 If $\sin \theta = \frac{1}{\sqrt{2}}$, write down the values of θ between −360° and 180°.

2. 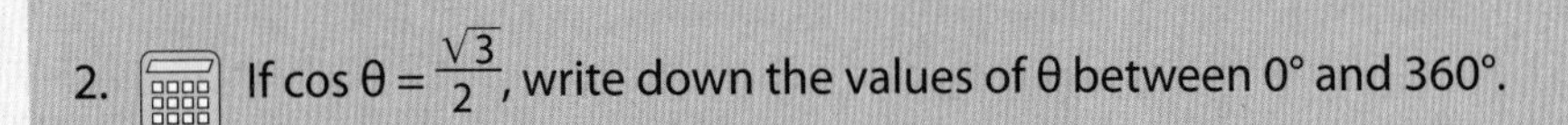 If $\cos \theta = \frac{\sqrt{3}}{2}$, write down the values of θ between 0° and 360°.

Arc, sector and segment

The length of an arc is a fraction of the circumference. The area of a sector is a fraction of the area of a circle.

Length of a circular arc

This can be expressed as a fraction of the circumference of a circle.

$$\text{Arc length} = \frac{\theta}{360°} \times 2\pi r$$

where θ is the angle subtended at the centre.

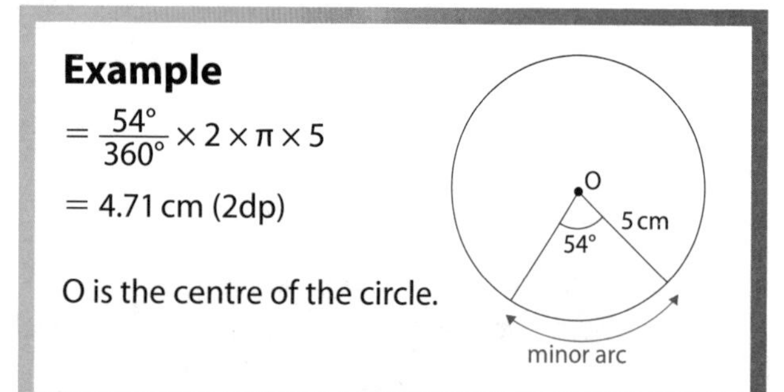

Example

$= \frac{54°}{360°} \times 2 \times \pi \times 5$

$= 4.71$ cm (2dp)

O is the centre of the circle.

Area of a sector

This can be expressed as the fraction of the area of a circle.

$$\text{Area of sector} = \frac{\theta}{360°} \times \pi r^2$$

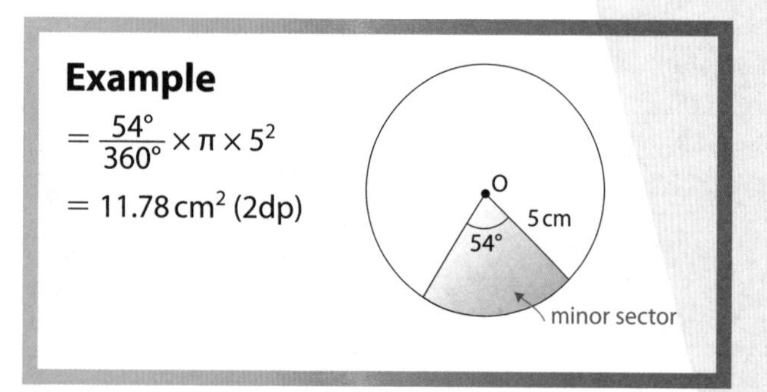

Example

$= \frac{54°}{360°} \times \pi \times 5^2$

$= 11.78\ \text{cm}^2$ (2dp)

Area of a general triangle

If we know the length of two sides of a triangle and the included angle, we can find the area.

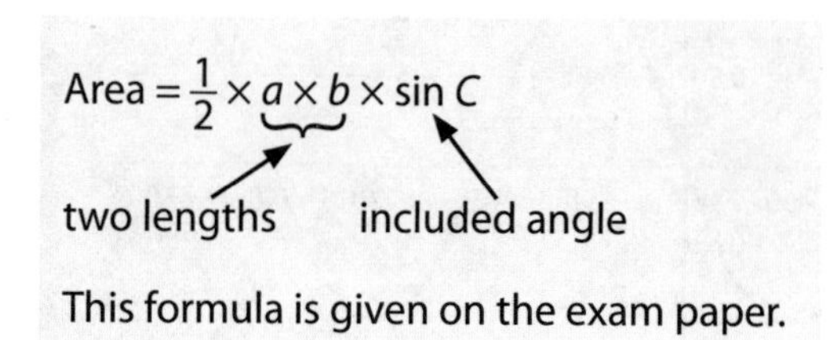

$$\text{Area} = \frac{1}{2} \times a \times b \times \sin C$$

This formula is given on the exam paper.

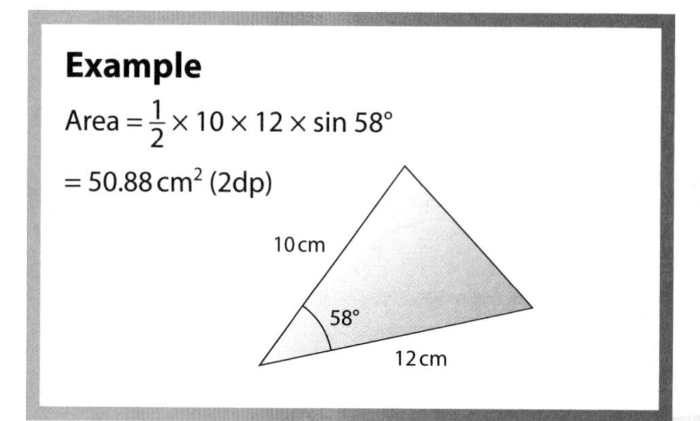

Example

$\text{Area} = \frac{1}{2} \times 10 \times 12 \times \sin 58°$

$= 50.88\ \text{cm}^2$ (2dp)

Area of a segment

This can be worked out in two stages:

1. Calculate the area of the sector and the area of the triangle.
2. Subtract the area of the triangle from the area of the sector.

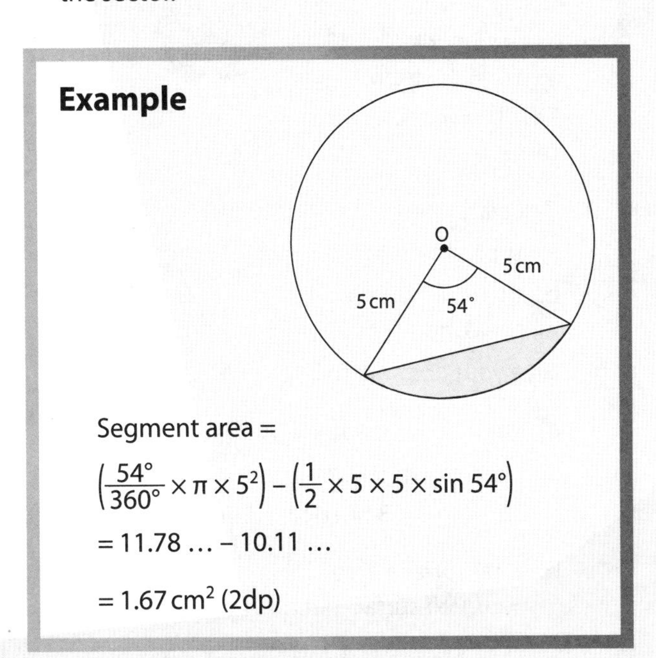

Example

Segment area =

$$\left(\frac{54°}{360°} \times \pi \times 5^2\right) - \left(\frac{1}{2} \times 5 \times 5 \times \sin 54°\right)$$

$= 11.78 \ldots - 10.11 \ldots$

$= 1.67\text{ cm}^2$ (2dp)

1 For this diagram, calculate:

a. the arc length (x)

b. the sector area

c. the area of the shaded segment

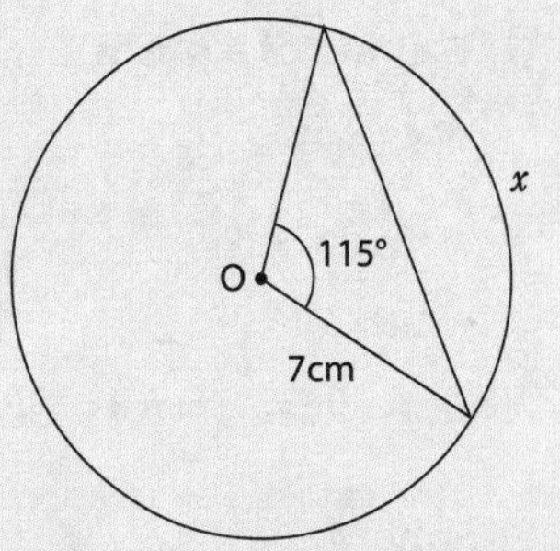

2 Decide whether this statement is true or false. You must show sufficient working out to justify your answer.

The area of the shaded segment is 9.03 cm^2.

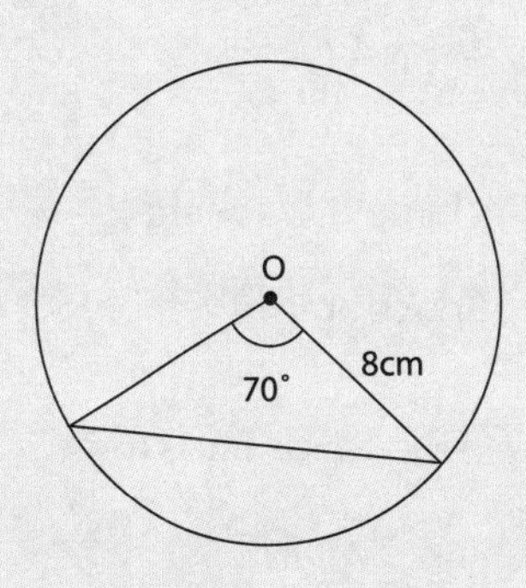

EXAM QUESTION

1. 

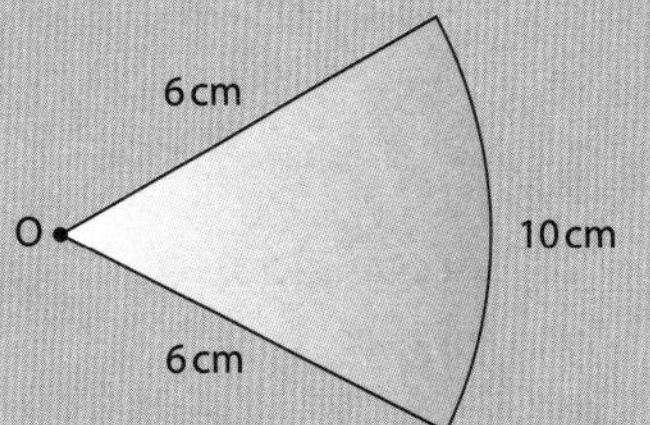

The diagram shows a sector of a circle, centre O, radius 6 cm.

The arc length of the sector is 10 cm.

Calculate the area of the sector in cm^2.

Surface area and volume

A prism is any solid that can be cut up into slices that are all the same shape. This is known as having a uniform cross-section.

Key formulae

Volume of a prism

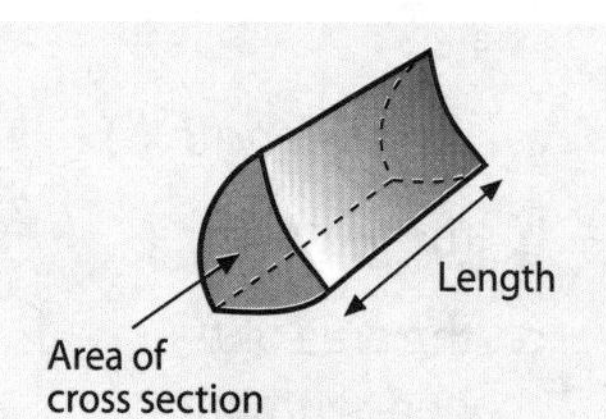

Volume = area of cross section × length $V = A \times l$

Sphere

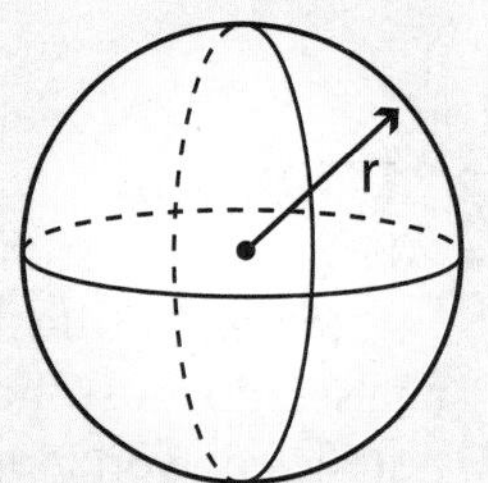

Volume of a sphere $= \frac{4}{3}\pi r^3$

Surface area of a sphere $= 4\pi r^2$

Pyramids and cones

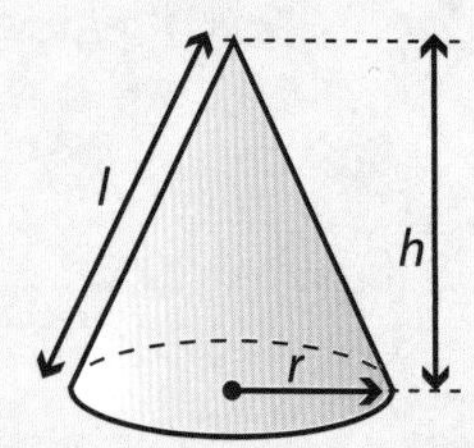

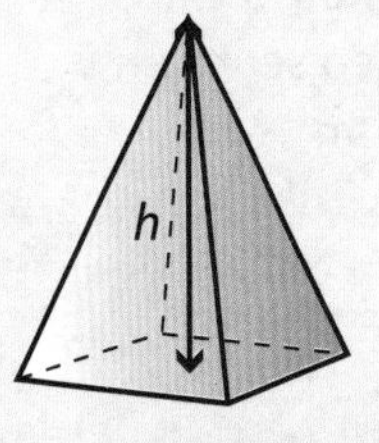

A cone is simply a pyramid with a circular base.

Volume of a pyramid $= \frac{1}{3} \times$ area of base × height

Volume of a cone $= \frac{1}{3} \times \pi \times r^2 \times$ height

Curved surface area of cone $= \pi r l$
(l is the slant height)

These formulae are given on the formula sheet.

Example

1. Find the volume of this cylinder:

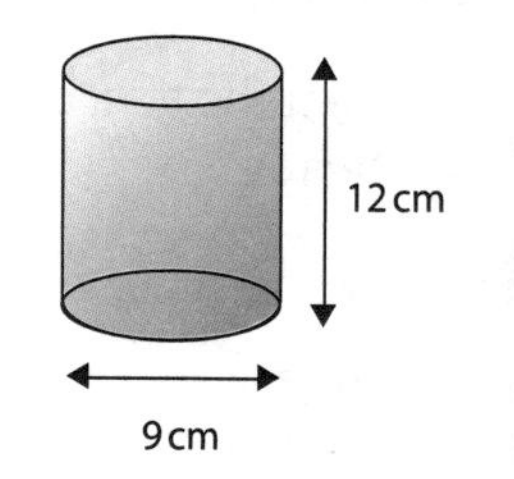

$V = \pi r^2 h$
$= \pi \times 4.5^2 \times 12$
$= 243\pi \text{ cm}^3$

This answer is left in terms of π.

2. Calculate
   a. the volume of the cone
   b. the total surface area of the cone

   Give your answers to 2dp.

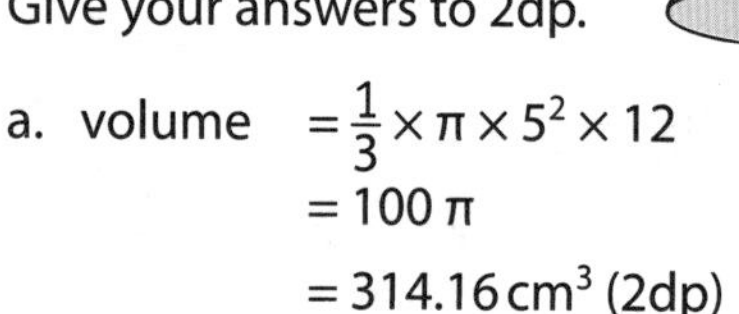

a. volume $= \frac{1}{3} \times \pi \times 5^2 \times 12$
$= 100\pi$
$= 314.16 \text{ cm}^3$ (2dp)

b. total surface area `
= curved surface area + area of circle

Slant height, $l = \sqrt{12^2 + 5^2}$
$= \sqrt{144 + 25}$
$= \sqrt{169}$
$= 13 \text{ cm}$

A $= \pi \times 5 \times 13 + \pi \times 5^2$
$= 65\pi + 25\pi$
$= 90\pi$
$= 282.74 \text{ cm}^2$ (2dp)
$= 282.74 \text{ cm}^2$

Dimensions

L = length L^2 = area L^3 = volume

- A formula with mixed dimensions is impossible, e.g. $L^2 + L^3$
- A dimension greater than 3 is impossible.

Example
If a, b and c represent lengths, what is the dimension of each of these expressions?

1. $\frac{abc}{3\pi} = L \times L \times L$ ∴ volume
2. $\sqrt{a^2 + b^2} = \sqrt{L^2 + L^2}$ = length
3. $a^2b + bc$ this has no dimensions since volume and area cannot be added together

PROGRESS CHECK

The volumes of the solids below have been calculated. Match the correct solid with the correct volume.

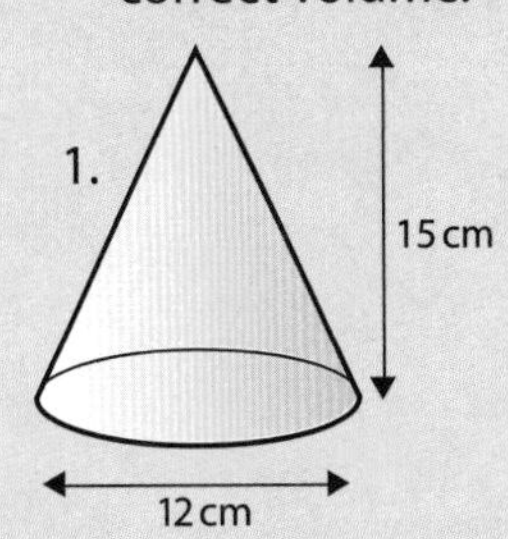

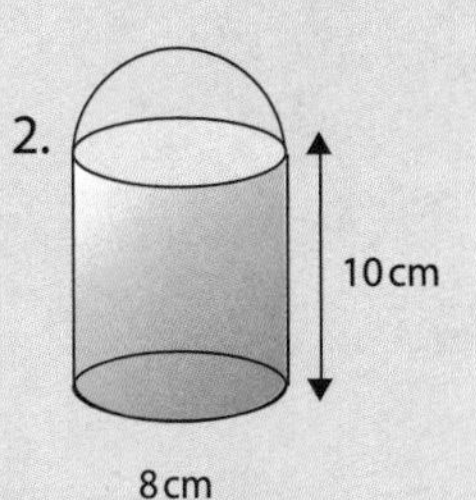

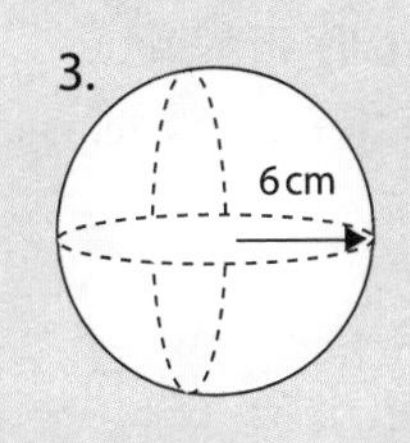

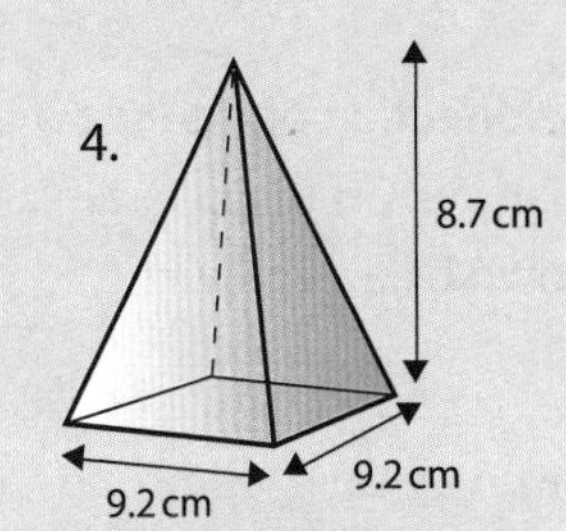

905 cm³ 245 cm³ 565 cm³ 637 cm³

EXAM QUESTION

1. Here are some expressions.

$\frac{1}{3}bd$	πc	$2d$	bcd	$c(b-d)$	$\frac{cd}{b}$	$\frac{\pi b^2 d}{c}$

The letters b, c and d represent lengths.
Three of the expressions could represent areas.
Tick the boxes underneath the three expressions that could represent areas.

2.  A cone has a volume of 15m³. The vertical height of the cone is 2.1m.
Calculate the radius of the base of the cone.

Vectors I

A vector is a quantity that has both distance and direction.

Vectors

- Four types of notation are used to represent vectors. The vector shown here can be referred to as

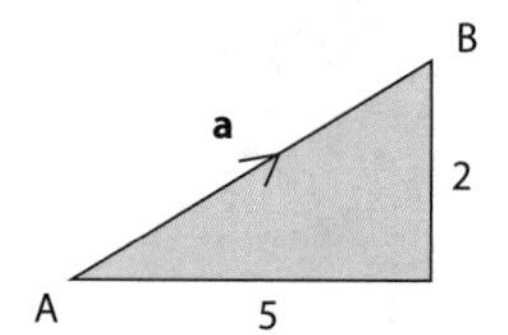

$\begin{pmatrix} 5 \\ 2 \end{pmatrix}$ or $\underline{a}$ or $\overrightarrow{AB}$ or **a**

- The direction of the vector is usually shown by an arrow. On the diagram above the vector $\overrightarrow{AB}$ is shown by an arrow.
- If $\overrightarrow{DE} = k\overrightarrow{AB}$ then $\overrightarrow{AB}$ and $\overrightarrow{DE}$ are parallel and the length of $\overrightarrow{DE}$ is k times the length of $\overrightarrow{AB}$.

$\overrightarrow{AB} = \begin{pmatrix} 5 \\ 2 \end{pmatrix}$

$\overrightarrow{DE} = \begin{pmatrix} 10 \\ 4 \end{pmatrix} = 2\begin{pmatrix} 5 \\ 2 \end{pmatrix}$

$\overrightarrow{DE} = 2\overrightarrow{AB}$

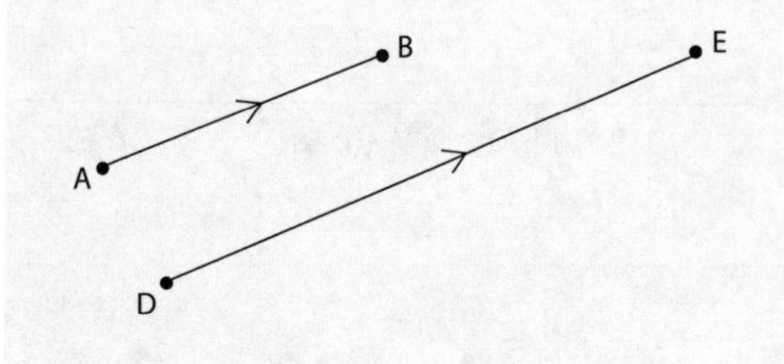

- If two vectors are equal, they are parallel and equal in length.

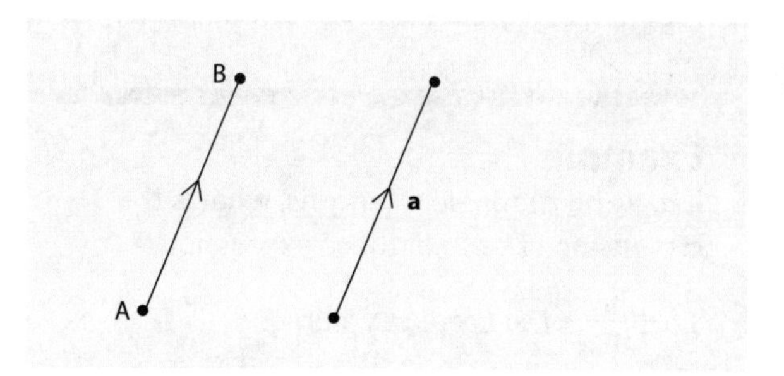

- If the vector $\mathbf{c} = \begin{pmatrix} 3 \\ 4 \end{pmatrix}$

then the vector $-\mathbf{c}$ is in the opposite direction to **c**.

$-\mathbf{c} = \begin{pmatrix} -3 \\ -4 \end{pmatrix}$

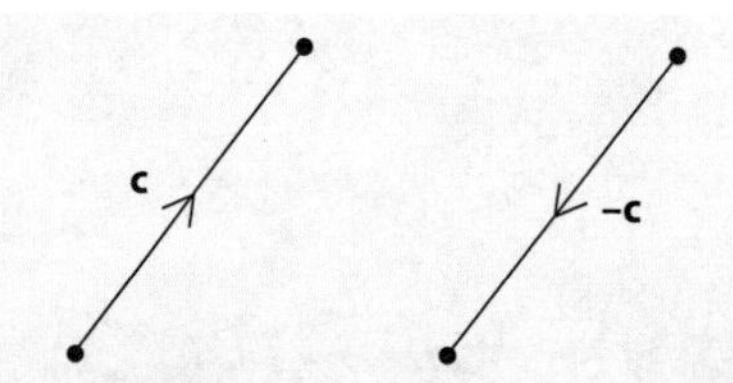

Magnitude of a vector

The magnitude of a vector is the length of the directed line segment representing it.

Pythagoras' theorem can be used to find the magnitude.

In general, the magnitude of a vector $\begin{pmatrix} x \\ y \end{pmatrix}$ is $\sqrt{x^2+y^2}$

The magnitude of AB

$(AB)^2 = 3^2 + 4^2$

$(AB)^2 = 9 + 16$

$(AB)^2 = 25$

$AB = \sqrt{25}$

$AB = 5$

The magnitude of AB is 5 units.

PROGRESS CHECK

1. The vector $\mathbf{b} = \begin{pmatrix} 4 \\ -2 \end{pmatrix}$. Draw vector **b**.
2. The vector $\mathbf{c} = \begin{pmatrix} -4 \\ 6 \end{pmatrix}$ and the vector $\mathbf{d} = \begin{pmatrix} -8 \\ 12 \end{pmatrix}$.

   Explain whether the vectors are parallel.
3. Work out the magnitude of the following vectors. Leave your answer in surd form, where possible.

   a. $\begin{pmatrix} 4 \\ 2 \end{pmatrix}$

   b. $\begin{pmatrix} -3 \\ 4 \end{pmatrix}$

   c. $\begin{pmatrix} -2 \\ 0 \end{pmatrix}$

   d. $\begin{pmatrix} 6 \\ -1 \end{pmatrix}$

EXAM QUESTION

1. The diagram is a sketch.
   A is the point (3,2)
   B is the point (7,7)

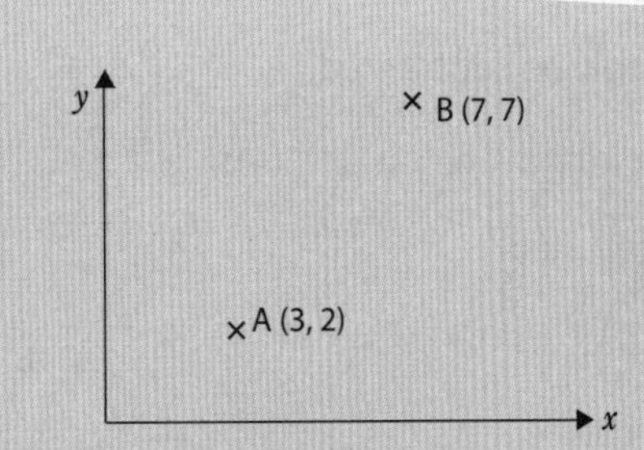

   a. Write down the vector $\overrightarrow{AB}$
      Write your answer as a column vector $\begin{pmatrix} x \\ y \end{pmatrix}$

   b. Write down the coordinates of point C such that $\overrightarrow{BC} = \begin{pmatrix} -2 \\ -3 \end{pmatrix}$

Vectors 2

Addition and subtraction of vectors

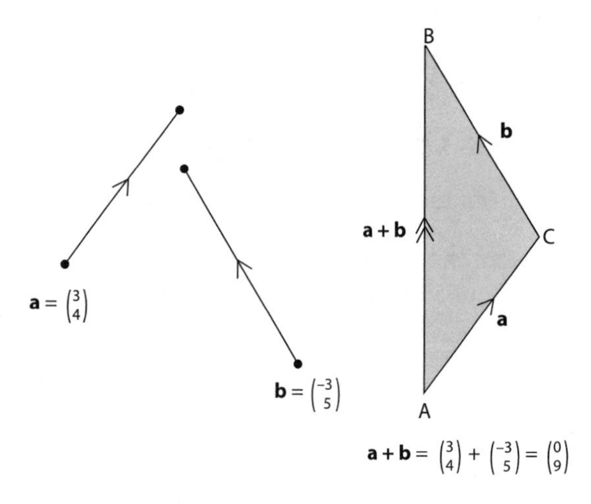

$\mathbf{a} + \mathbf{b} = \begin{pmatrix}3\\4\end{pmatrix} + \begin{pmatrix}-3\\5\end{pmatrix} = \begin{pmatrix}0\\9\end{pmatrix}$

- The **resultant** of two vectors is found by adding them.
- Vectors must always be added end to end so that the arrows follow on from each other.
- A resultant is usually labelled with a double arrow.

The triangle above shows the triangle law of addition.

To take the route directly from A to B, is equivalent to travelling via C, hence we can represent $\overrightarrow{AB}$ as $\mathbf{a} + \mathbf{b}$.

Vectors can also be subtracted.

$\mathbf{a} - \mathbf{b}$ can be interpreted as $\mathbf{a} + (-\mathbf{b})$.

$\mathbf{a} + (-\mathbf{b}) = \begin{pmatrix}3\\4\end{pmatrix} + \begin{pmatrix}3\\-5\end{pmatrix}$

$\therefore\ \mathbf{a} - \mathbf{b} = \begin{pmatrix}6\\-1\end{pmatrix}$

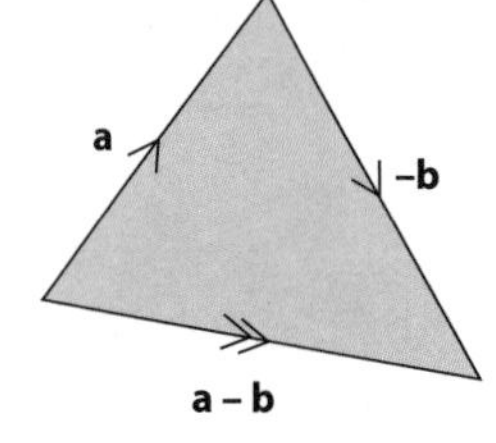

Position vectors

- The position vector of a point Q is the vector $\overrightarrow{OQ}$, where O is the origin.

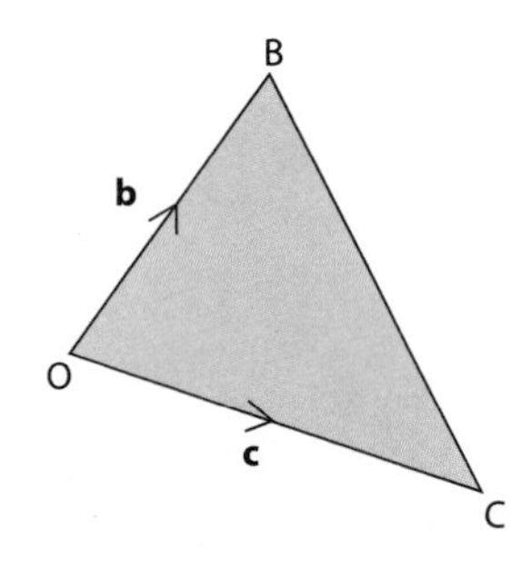

In the diagram the position vectors of B and C are **b** and **c** respectively. Using this notation:

$\overrightarrow{BC} = -\mathbf{b} + \mathbf{c}$ or $\mathbf{c} - \mathbf{b}$

Example

OAB is a triangle. Given that $\overrightarrow{OA} = \mathbf{a}$, $\overrightarrow{OB} = \mathbf{b}$ and that N splits $\overrightarrow{AB}$ in the ratio 1 : 2, find in terms of **a** and **b** the vectors:

i. $\overrightarrow{AB}$ ii) $\overrightarrow{ON}$

i. $\overrightarrow{AB} = \overrightarrow{OA} + \overrightarrow{OB}$ (go from A to B via O)

$= -\mathbf{a} + \mathbf{b}$

ii. $\overrightarrow{ON} = \overrightarrow{AO} + \overrightarrow{AN}$ $(\overrightarrow{AN} = \frac{1}{3}\overrightarrow{AB})$

$= \mathbf{a} + \frac{1}{3}(-\mathbf{a} + \mathbf{b})$

$= \mathbf{a} - \frac{1}{3}\mathbf{a} + \frac{1}{3}\mathbf{b}$

$= \frac{2}{3}\mathbf{a} + \frac{1}{3}\mathbf{b}$

$= \frac{1}{3}(2\mathbf{a} + \mathbf{b})$

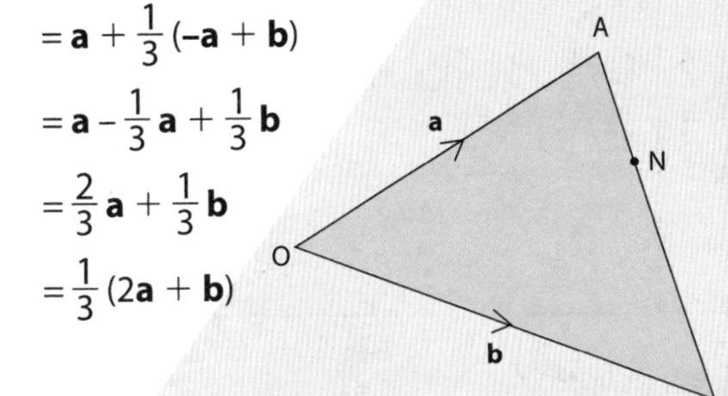

1. OABC is a parallelogram.
AB is parallel to OC.
OA is parallel to CB.

$\overrightarrow{OA} = \mathbf{a}$
$\overrightarrow{OC} = \mathbf{c}$

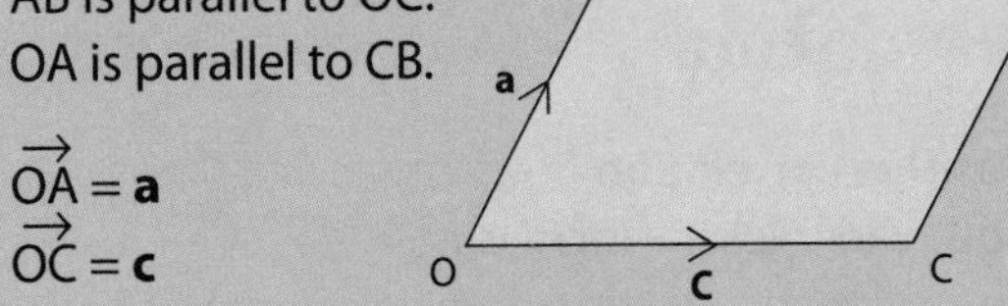

a. Express in terms of **a** and **c**

i. $\overrightarrow{AC}$

ii. $\overrightarrow{BO}$

b. If N is the midpoint of $\overrightarrow{AC}$.

Express $\overrightarrow{ON}$ in terms of **a** and **c**.

EXAM QUESTION

ABCDEF is a regular hexagon.

Given that $\overrightarrow{OB} = \mathbf{a}$ and $\overrightarrow{OC} = \mathbf{b}$

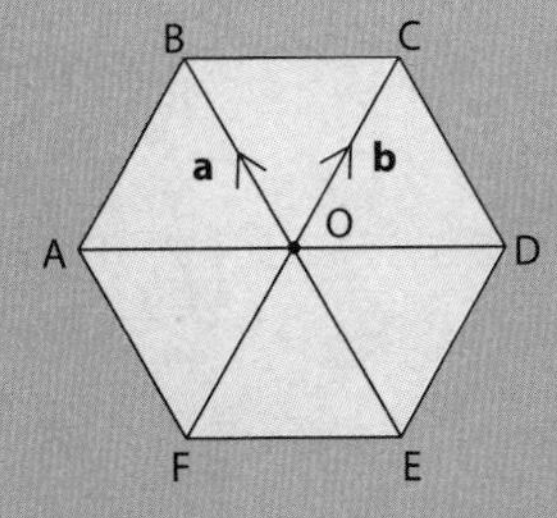

1. Find in terms of **a** and **b** the vectors:

a. $\overrightarrow{BC}$

b. $\overrightarrow{AD}$

2. Write down the vector $\overrightarrow{FE}$.

3. What geometrical fact is exhibited by the vectors $\overrightarrow{FE}$ and $\overrightarrow{AD}$?

Scatter diagrams and correlation

Scatter diagrams are used to show two sets of data at the same time. They are important because they show the correlation (connection) between the sets.

There are three types of correlation

- **Positive correlation**
  Both variables are increasing.
  The taller you are, the more you probably weigh.

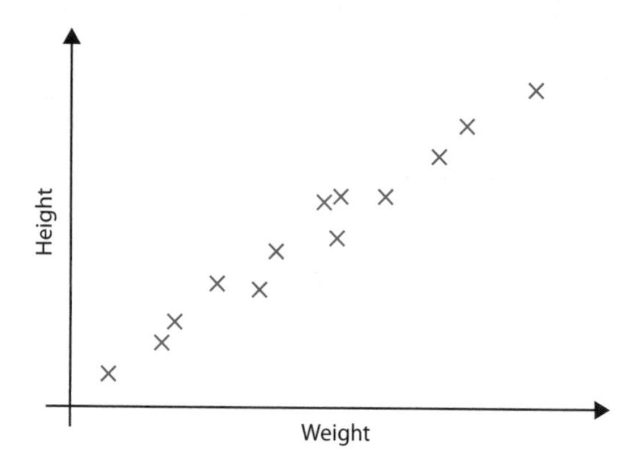

- **Negative correlation**
  As one variable increases the other decreases.
  As the temperature increases, the sale of woollen hats probably decreases.

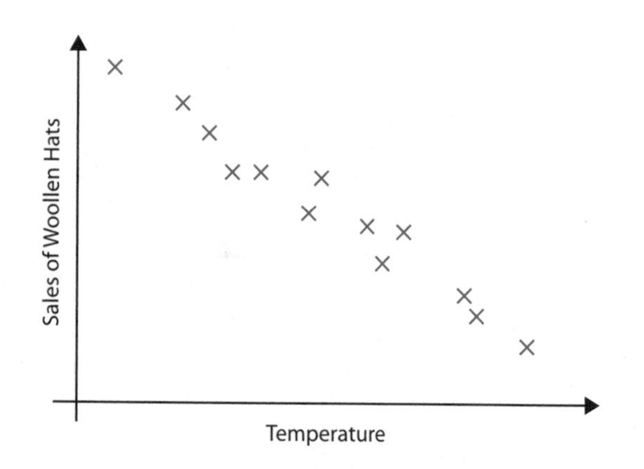

- **Zero correlation**
  Little or no correlation between the variables.
  No connection between your height and your mathematical ability.

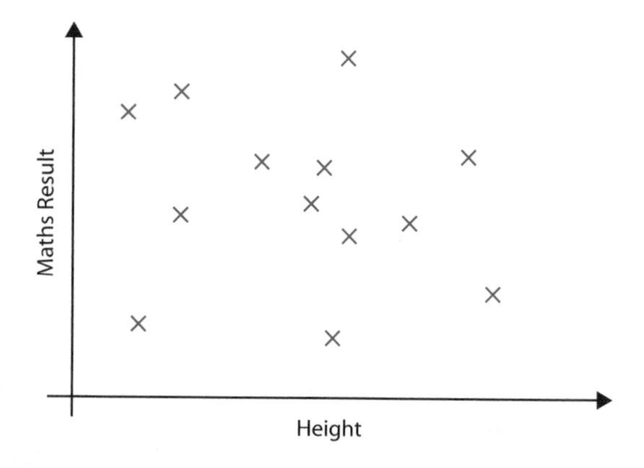

PROGRESS CHECK

Decide whether these statements are **true** or **false**.

1. There is a positive correlation between the weight of a book and the number of pages.
2. There is no correlation between the height you climb up a mountain and the temperature.
3. There is a negative correlation between the age of a used car and its value.
4. There is a positive correlation between the height of some students and the size of their feet.
5. There is no correlation between the weight of some students and their History GCSE results.

Line of best fit

- The line goes as close as possible to all the points.
- There is roughly an equal number of points above the line and below it.

Example

The scatter diagram shows the Science and Maths percentages scored by some students.

- The line of best fit goes in the direction of the data.
- We can estimate that a student with a Science percentage of 30 would get a Maths percentage of about 17.
- We can estimate that a student with a Maths percentage of 50 would get about 54% in Science.
- This shows how the line of best fit can be used to estimate results.

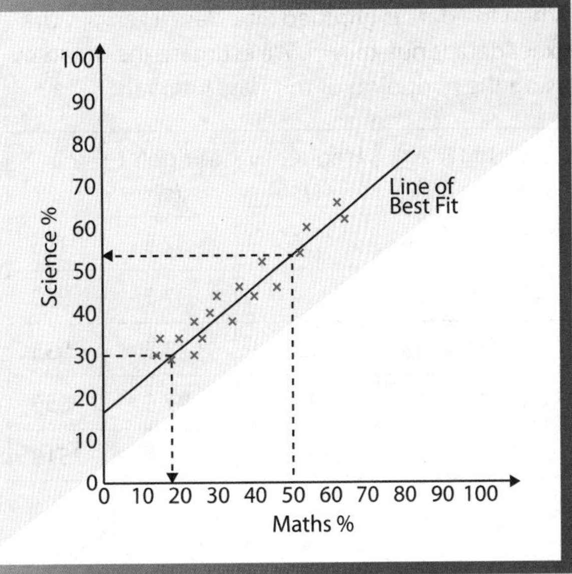

EXAM QUESTION

1. The scatter diagram shows the age of some cars and their values.

   a. Draw a line of best fit on the diagram.

   b. Use your line of best fit to estimate the age of a car, when its value is £5 000.

   c. Use your line of best fit to estimate the value of a $3\frac{1}{2}$ year old car

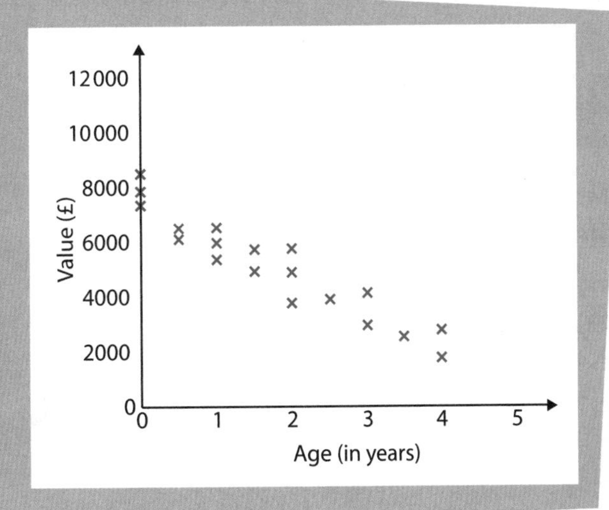

Averages

You need to understand the difference between mean, median and mode and be able to find them in the examination.

Averages of continuous data

When the data is grouped into class intervals, the exact data is not known. We estimate the mean by using the midpoints of the class intervals.

Weight (W kg)	Frequency (f)	Midpoint (x)	fx
$30 \leqslant W < 35$	6	32.5	195
$35 \leqslant W < 40$	14	37.5	525
$40 \leqslant W < 45$	22	42.5	935
$45 \leqslant W < 50$	18	47.5	855
	60		2510

$\bar{x} = \dfrac{\Sigma fx}{\Sigma f}$ — Σ means the sum of

$\bar{x} = \dfrac{2510}{60}$ — f represents the frequency

$\bar{x} = 41.8\dot{3}$ (2dp) — $\bar{x}$ represents the mean

Add in 2 extra columns – one for the midpoint and one for fx.

Modal class is $40 \leqslant W < 45$.
This class interval has the highest frequency.

The example above is a very common question at GCSE, usually worth 4 marks.

Finding the median

To find the class interval containing the median, firstly find the position of the median.

$$= \frac{\Sigma(f+1)}{2}$$

$$= \frac{(60+1)}{2} = 30.5.$$

The median lies between the 30th and 31st value. The 30th and 31st value are in the class interval:

$40 \leqslant W < 45$

The class interval in which the median lies is $40 \leqslant W < 45$.

Moving averages

- Used to smooth out the changes in a set of data that varies over a period of time.
- A four-point moving average uses four data items in each calculation, a three-point moving average uses three, and so on.
- Used to show the trend in a set of data.
- Can be used to draw a trend line on a time series graph.

Example

Find the four-point moving average for the following data:

3 2 0 1 4 6

Average for 1st 4 data points
$(3 + 2 + 0 + 1) \div 4 = 1.5$

Average for data points 2 to 5
$(2 + 0 + 1 + 4) \div 4 = 1.75$

Average for data points 3 to 6
$(0 + 1 + 4 + 6) \div 4 = 2.75$

1. The heights, h cm, of some students are shown in the table.

Height (cm)	Frequency
$140 \leq h < 145$	4
$145 \leq h < 150$	9
$150 \leq h < 155$	15
$155 \leq h < 160$	6

Calculate an estimate for the mean of this data.

2. Work out the three-point moving averages for these data:

2, 1, 3, 4, 5, 6

EXAM QUESTION

1. The table shows information about the number of hours that 50 children watched television for last week:

Work out an estimate for the mean number of hours the children watched television.

Number of hours (h)	Frequency
$0 \leq h < 2$	3
$2 \leq h < 4$	6
$4 \leq h < 6$	22
$6 \leq h < 8$	13
$8 \leq h < 10$	6

Cumulative frequency graphs

Cumulative frequency graphs are useful for finding the median and spread of grouped data

With a cumulative frequency graph it is possible to estimate the median of grouped data and the interquartile range.

Example

The table shows the marks of 94 students in a Mathematics exam.

a. Complete the cumulative frequency table for this data.

b. Draw a cumulative frequency graph for this data.
To do this we must plot the top value of each class interval on the x-axis and the cumulative frequency on the y-axis.
Plot (20, 2) (30, 8) (40, 18) …
Join the points with a smooth curve.
Since no students had less than zero marks, the graph starts at (0, 0).

Mark	Frequency	Mark	Cumulative Frequency
0–20	2	$\leqslant 20$	2 (2)
21–30	6	$\leqslant 30$	8 (2+6)
31–40	10	$\leqslant 40$	18 (8+10)
41–50	17	$\leqslant 50$	35 (18+17)
51–60	24	$\leqslant 60$	59 (35+24)
61–70	17	$\leqslant 70$	76 (59+17)
71–80	11	$\leqslant 80$	87 (76+11)
81–90	4	$\leqslant 90$	91 (87+4)
91–100	3	$\leqslant 100$	94 (91+3)

Cumulative frequency is a running total of all the frequencies.

This means that 87 students had a score of 80 or less

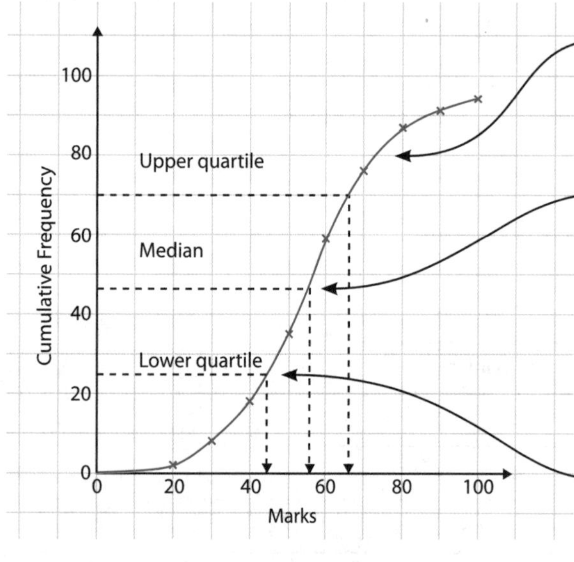

The **upper quartile** is three quarters of the way into the distribution

$\frac{3}{4} \times 94 = 70.5$

Read across from 70.5 Upper quartile $\approx$ 66.5 marks

The **median** splits the data into two halves – the lower 50% and the upper 50%.

Median $= \frac{1}{2} \times$ cumulative frequency

$= \frac{1}{2} \times 94 = 47$

Read across from 47 and down to the horizontal axis.
Median $\approx$ 56 marks

The **lower quartile** is the value one quarter into the distribution

$\frac{1}{4} \times 94 = 23.5$

Read across from 23.5 Lower quartile $\approx$ 44 marks

Interquartile range

= upper quartile – lower quartile

= 66.5 − 44 = 22.5 marks

- A large interquartile range indicates that the 'middle half' of the data is widely spread about the median.
- A small interquartile range indicates that the 'middle half' of the data is concentrated about the median.

Box plots

- Sometimes known as box and whisker diagrams.
- Cumulative frequency graphs are not easy to compare; a box plot shows the Interquartile range as a box.

Example

The box plot of the cumulative frequency graph opposite would look like this:

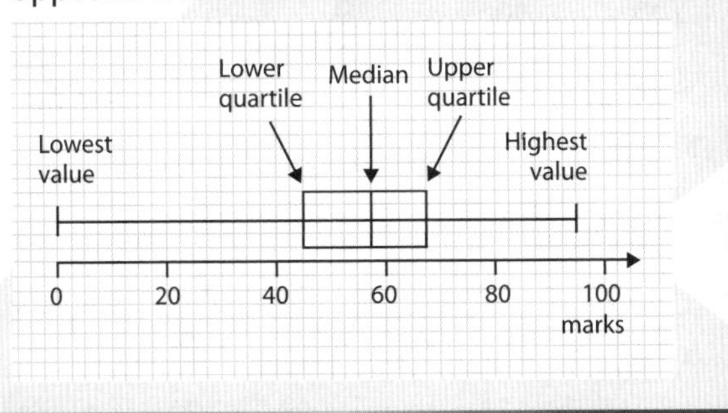

PROGRESS CHECK

1. Students in 9A and 9B took the same test. Their results are used to draw the following box plots.

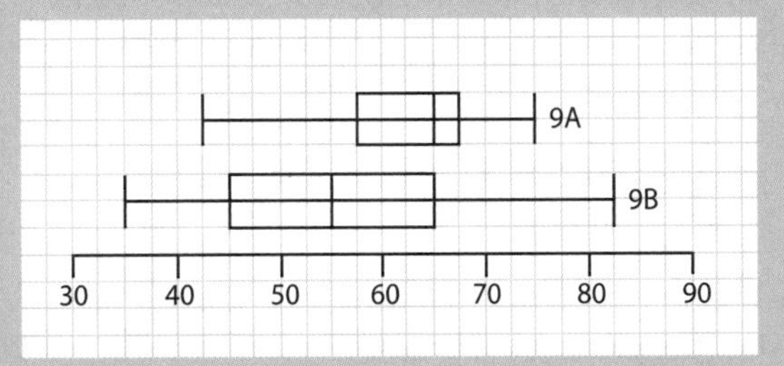

a. In which class was the student who scored the highest mark?

b. In which class did the students perform better in the test? You must give a reason for your answer.

EXAM QUESTION

1. The times in minutes to finish an assault course are listed in order 8, 12, 12, 13, 15, 17, 22, 23, 23, 27, 29

a. From the data, find:

i. the lower quartile

ii. the interquartile range

b. Draw a box plot for this data.

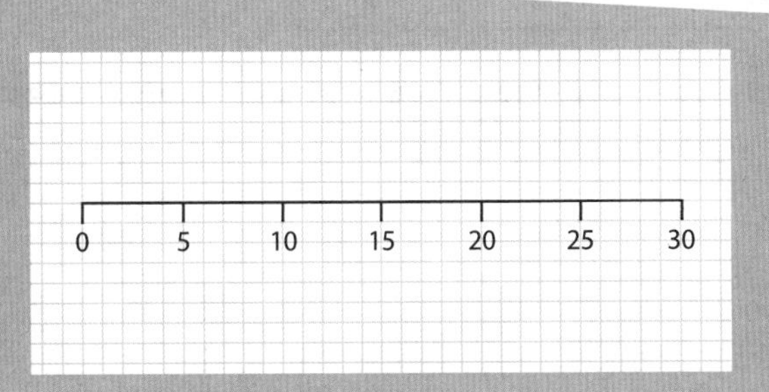

Histograms

Histograms are similar to bar charts except their bars can be different widths. The area of each 'bar' represents the frequency.

Drawing histograms

- When the widths of the bars are different the vertical axis is known as the frequency density.

$$\text{frequency density} = \frac{\text{frequency}}{\text{class width}}$$

- The areas of the rectangles are equal to the frequencies they represent.

Example

The table shows the time in seconds it takes people to swim 100 metres. Draw a histogram of this information.

Time, t (seconds)	Frequency
$100 < t \leqslant 110$	2
$110 < t \leqslant 140$	24
$140 < t \leqslant 160$	42
$160 < t \leqslant 200$	50
$200 < t \leqslant 220$	24
$220 < t \leqslant 300$	20

- To draw a histogram you need to calculate the frequency densities. Add an extra column to the table.

The table should now look like this.

Time, t (seconds)	Frequency	Frequency density
$100 < t \leqslant 110$	2	$2 \div 10 = 0.2$
$110 < t \leqslant 140$	24	$24 \div 30 = 0.8$
$140 < t \leqslant 160$	42	$42 \div 20 = 2.1$
$160 < t \leqslant 200$	50	$50 \div 40 = 1.25$
$200 < t \leqslant 220$	24	$24 \div 20 = 1.2$
$220 < t \leqslant 300$	20	$20 \div 80 = 0.25$

- Draw on graph paper – make sure there are no gaps!

The graph should now look like this.

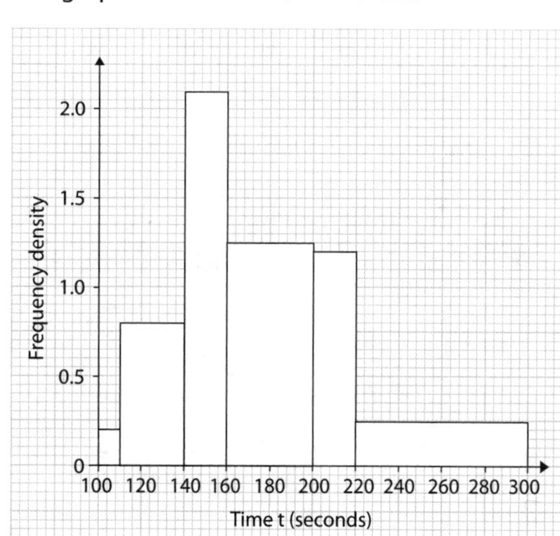

- Sometimes you will be asked to read from a histogram. In this case rearrange the formula for frequency density.

frequency =
frequency density × class width

PROGRESS CHECK

1. The table gives some information about the ages of some participants in a charity walk.

Age (x) in years	Frequency
$0 < x \leq 15$	30
$15 < x \leq 25$	46
$25 < x \leq 40$	45
$40 < x \leq 60$	25

On graph paper draw a histogram to represent this information.

EXAM QUESTION

The table and histogram give information about the distance (d km) travelled to work by some employees.

Distance (km)	Frequency	Frequency density
$0 < d \leq 15$	12	
$15 < d \leq 25$	....	
$25 < d \leq 30$	36	
$30 < d \leq 45$	....	
$45 < d \leq 55$	20	

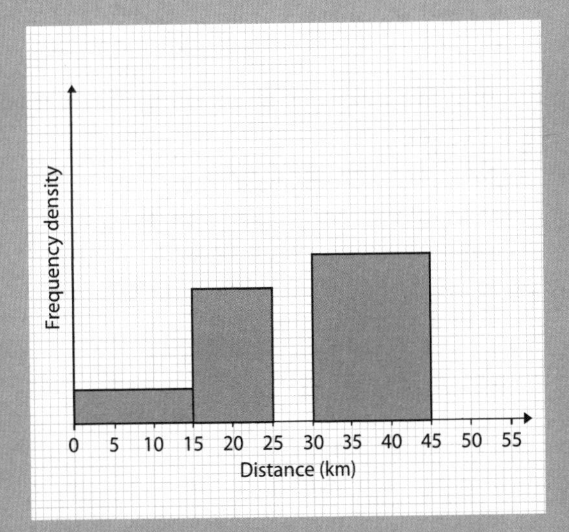

1. Use the information in the histogram to complete the table.

2. Use the table to complete the histogram.

Probability

Probability is the chance or likelihood that something will happen. All probabilities lie between 0 and 1. Probabilities must be written as either a fraction, decimal or percentage.

Tree diagrams

Tree diagrams are used to show the possible outcomes of two or more events. There are two rules you need to know first.

- **The OR rule**
  If two or more events are mutually exclusive the probability of A or B happening is found by adding the probabilities.

  P(A or B) = P(A) + P(B)

  (This also works for more than two outcomes)

- **The AND rule**
  If two or more events are independent, the probability of A and B and C happening together is found by multiplying the separate probabilities

  P(A and B and C …) = P(A) × P(B) × P(C) …

Example

A bag contains 3 red and 4 blue counters. A counter is taken from the bag at random, its colour is noted and then it is replaced in the bag. A second counter is then taken out of the bag. Draw a tree diagram to illustrate this information.

- Remember that the branches leaving each point on the tree add up to 1.

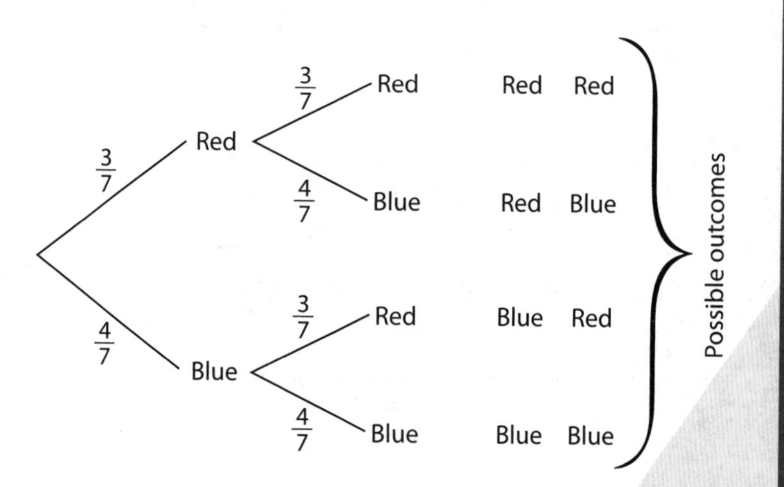

Example

Work out the probability of:

i. picking two blues
ii. picking one of either colour

i. P(two blues) = P(B) × P(B)

$$= \frac{4}{7} \times \frac{4}{7} = \frac{16}{49}$$

■ Remember to multiply along the branches.

ii. P(one of either colour)

P(B) × P(R)

$$= \frac{4}{7} \times \frac{3}{7} = \frac{12}{49}$$

OR (**OR** means **add**)

P(R) × P(B)

$$= \frac{3}{7} \times \frac{4}{7} = \frac{12}{49}$$

P(one of either colour)

$$= \frac{12}{49} + \frac{12}{49} = \frac{24}{49}$$

■ Remember to include all possibilities.

Charlotte has a biased die. The probability of getting a three is 0.4. She rolls the die twice.

1. Complete the tree diagram.

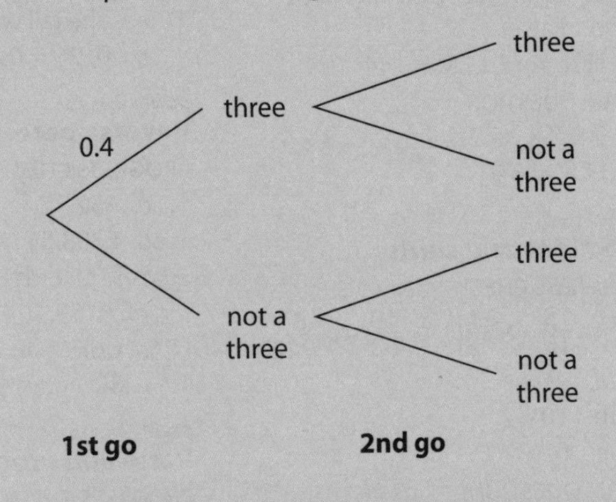

2. Work out the probability that she gets:

   a. two threes

   b. exactly one three

1. A bag contains 3 red, 4 blue and 2 green beads. A bead is picked out of the bag at random and its colour noted. It is not replaced in the bag. A second bead is picked out at random. Work out the probability that two different coloured beads are chosen.

Answers

Day 1

pages 4–5

Prime factors, HCF and LCM

PROGRESS CHECK

1. a. $50 = 2 \times 5^2$
   b. $360 = 2^3 \times 3^2 \times 5$
   c. $16 = 2^4$
2. a. False b. True c. True
   d. False
3. HCF = 12 LCM = 144

EXAM QUESTION

1. $2 \times 2 \times 2 \times 3 \times 5$ or $2^3 \times 3 \times 5$
2. LCM = 840

pages 6–7

Fractions and surds

PROGRESS CHECK

1. a. $\frac{13}{15}$ b. $\frac{11}{21}$ c. $\frac{10}{63}$ d. $\frac{81}{242}$
2. a. $2\sqrt{6}$
   b. $10\sqrt{2}$
   c. $6\sqrt{3}$

EXAM QUESTION

1. $-\frac{8}{3}$
2. $\frac{1}{\sqrt{2}} \times \frac{\sqrt{2}}{\sqrt{2}} = \frac{\sqrt{2}}{2}$

pages 8–9

Percentages

PROGRESS CHECK

1. a. 10kg b. £600
   c. £3 d. 252g
2. a. 32% b. 23%
   c. 75% d. 84%
3. £180
4. £28.75
5. 77.1%

EXAM QUESTION

1. £703.83
2. 80%

pages 10–11

Repeated percentage change

PROGRESS CHECK

1. 51.7% (3sf)
2. 29.6% (3sf)
3. £5518.28
4. £121 856

EXAM QUESTION

1. a. The fall will be less than 100%
   b. $0.9^2 = 0.81$ c. £10935

pages 12–13

Reverse percentage problems

PROGRESS CHECK

1. a. £52.77 b. £106.38
   c. £208.51 d. £446.81
2. Yes, Joseph is correct since $\frac{60}{0.85}$ =£70.59.

EXAM QUESTION

1. £250

pages 14–15

Ratio and Proportion

PROGRESS CHECK

1. £20 : £40 : £100
2. £35.28
3. 3 : 4
4. 280g
5. 3 days

EXAM QUESTIONS

1. Amy £48
2. 250g of flour
   150g of sugar
   125g of margarine
   100g of chocolate chips
   5 eggs

Day 2

pages 16–17

Indices

PROGRESS CHECK

1. a. 6^8 b. 12^{13} c. 7^{24}
   d. 5^6 e. 2^5
2. a. $6b^{10}$ b. $2b^{-16}$ c. $9b^8$
   d. $6b^{14}$ e. $\frac{1}{25x^4y^6}$

EXAM QUESTIONS

1. a. 1 b. $\frac{1}{49}$ c. 4 d. $\frac{1}{9}$

pages 18–19

Standard index form

PROGRESS CHECK

1. a. 6.4×10^4 b. 2.71×10^5
   c. 4.6×10^{-4} d. 7.4×10^{-8}
2. a. 1.2×10^{11} b. 2×10^{-1}
   c. 3.5×10^{16}
3. a. 1.4375×10^{18} b. 5.48×10^{19}

EXAM QUESTION

1. a. 4×10^7 b. 0.00006
2. 2.4

pages 20–21

Proportionality

PROGRESS CHECK

1. a. $y = kx$ b. $y = \frac{k}{\sqrt[3]{x}}$
   c. $y = \frac{k}{x}$ d. $y = kx^3$
2. $a = k\sqrt{x}$
   $8 = k\sqrt{4}$
   $\therefore k = 4$
   $a = \pm 4\sqrt{x}$
   $64 = 4\sqrt{x}$
   $x = 256$

EXAM QUESTION

1. a. $y = \frac{90}{x^2}$ b. $y = \frac{90}{4}$ $y = 22.5$
   c. $x = 3.87$

pages 22–23

Upper and lower bounds

PROGRESS CHECK

1. A – upper bound
   E – lower bound
2. Upper bound = $0.22\dot{6}$
   Lower bound = $0.2\dot{1}\dot{8}$

EXAM QUESTION

1. 0.0437 (3sf)

pages 24–25

Formulae and Expressions

PROGRESS CHECK

1. a. $8a - 7b$ b. $4a^2 - 8b^2$
   c. $2xy + 2xy^2$
2. a. $-\frac{31}{5} = -6\frac{1}{5}$ b. 4.36 c. 9
3. $u = \pm\sqrt{v^2 - 2as}$
4. $p = \frac{-(t + vq)}{q - 1}$ or $\frac{(t + vq)}{1 - q}$

EXAM QUESTION

1. a. Josh is correct since $3x^2$ means $3 \times x^2$, this gives $3 \times 2^2 = 12$
   b. 36
   c. $X = \pm\sqrt{\frac{y}{4}} + 1 = \frac{\sqrt{y}}{2} + 1$

pages 26–27

Brackets and Factorisation

PROGRESS CHECK

1. a. $x^2 + x - 6$ b. $6x - 8$
   c. $4x^2 - 12x$ d. $x^2 - 6x + 9$
2. a. $4x(x + 2)$ b. $6x(2y - x)$
   c. $3ab(a + 2b)$
3. a. $(x + 2)(x + 2)$ b. $(x - 2)(x - 3)$
   c. $(x + 1)(x - 5)$

EXAM QUESTION

1. a. i. $3t^2 - 4t$ ii. $6x + 4$
   b. i. $y(y + 1)$ ii. $5pq(p - 2q)$
   iii. $(a + b)(a + b + 4)$
   iv. $(x - 2)(x - 3)$

pages 28–29

Equations 1

PROGRESS CHECK

1. $x = 8$
2. $x = -5$
3. $x = -\frac{1}{2}$
4. $x = -3.25$
5. $x = -\frac{1}{2}$
6. $x = 17$

EXAM QUESTION

1. a. $x = 2.4$ b. $x = -2.5$ c. $y = 2$
2. a. $x = -\frac{1}{5}$ b. $x = -4\frac{1}{3}$

Day 3

pages 30–31

Equations 2

PROGRESS CHECK

1. $x = 5.5$ cm, shortest length: $2x - 5 = 6$ cm
2. $k + 3 = \frac{1}{3}$ $k = -\frac{8}{3}$
3. $4^{2k} = 4^3$ $2k = 3$ $k = \frac{3}{2}$
4. $2^{3k-1} = 2^6$ $3k - 1 = 6$ $k = \frac{7}{3}$

EXAM QUESTION

1. a. $x + 30° + 2x + x + 50° + x + 10° = 360°$ $5x + 90° = 360°$
   b. $x = 54°$ smallest angle = 64°

pages 32–33

Solving quadratic equations 1

PROGRESS CHECK

1. a. $(x - 2)(x - 4) = 0$ $x = 2, x = 4$
   b. $(x+4)(x + 1) = 0$ $x = -4, x = -1$
   c. $(x + 2)(x - 6) = 0$ $x = -2, x = 6$
2. a. $x = -5, x = 3$
   b. $x = -\frac{1}{2}, x = -2$

EXAM QUESTION

1. a. $(2x + 5)(x + 3) - (x + 1)^2$
   $(2x^2 + 11x + 15) - (x^2 + 2x + 1)$
   Area $= x^2 + 9x + 14$
   Since the shaded area = 45 cm^2
   $45 = x^2 + 9x + 14$
   So $x^2 + 9x + 14 - 45 = 0$
   $x^2 + 9x - 31 = 0$
   b. i. $x = \frac{-b \pm \sqrt{b^2 - 4ac}}{2a}$
   $x = \frac{-9 \pm \sqrt{9^2 - (4 \times 1 \times -31)}}{2 \times 1}$
   $x = 2.66$cm (3sf)
   ii. perimeter $3.66... \times 4 =$ 14.6cm (3sf)

pages 34–35

Solving quadratic equations 2

PROGRESS CHECK

1. 3.3 (1dp)
2. a. $(x + 4)^2 - 14$ $\therefore a = 4, b = -14$
   b. -14

EXAM QUESTION

1. $x = 2.7$
2. $p = -2,\ q = 3$

pages 36–37

Simultaneous linear equations

PROGRESS CHECK

1. a. $b = -4.5$ $a = 4$ b. $p = -3, r = 4$
   c. $x = 2, y = 3$
2. $x = 2,\ y = 4$

EXAM QUESTION

1. $a = 3$ $b = -2$

pages 38–39

Solving linear and quadratic equations simultaneously

PROGRESS CHECK

1. a. $x = -1, y = 3$ $x = 2, y = 6$
   b. $x = 3, y = 4$ $x = -4, y = -3$
2. a. This is where the quadratic graph $y = x^2 + 2$ intersects the straight line graph $y = x + 4$. Their points of intersection are (–1, 3) and (2, 6).
   b. This is where the circle $x^2 + y^2 = 25$ intersects with the straight line $y = x + 1$. Their points of intersection are (3, 4) and (–4, –3).

EXAM QUESTION

1. a. $x = 1,\ y = 5$
   b. $x = -3.4, y = -3.8$

Day 4

pages 40–41

Algebraic fractions

PROGRESS CHECK

1. a. no b. yes c. yes
2. a. $\frac{5x - 1}{(x + 1)(x - 1)}$ b. $\frac{2aq}{3c}$
   c. $\frac{40(a + b)}{3}$

EXAM QUESTION

1. $\frac{8(x + 2)}{(x + 2)(x - 2)} = \frac{8}{(x - 2)}$
2. $\frac{x^2(6 + x)}{x^2 - 36} = \frac{x^2(6 + x)}{(x + 6)(x - 6)} = \frac{x^2}{(x - 6)}$
3. $\frac{2x^2 + 7x - 15}{x^2 + 3x - 10} = \frac{(2x - 3)(x + 5)}{(x + 5)(x - 2)}$
   $= \frac{2x - 3}{x - 2}$

pages 42–43

Sequences

PROGRESS CHECK

1. a. $4n+1$ b. $2-n$ c. $2n$ d. $3n+2$ e. $5n-1$
2. a. $3n+4$ b. $\frac{1}{2n+1}$ c. $3n-2$

EXAM QUESTION

1. $2n+3$

pages 44–45

Inequalities

PROGRESS CHECK

1. a. $x < 2.2$ b. $\frac{4}{3} \leqslant x < 3$ c. $x > -\frac{9}{5}$
2.

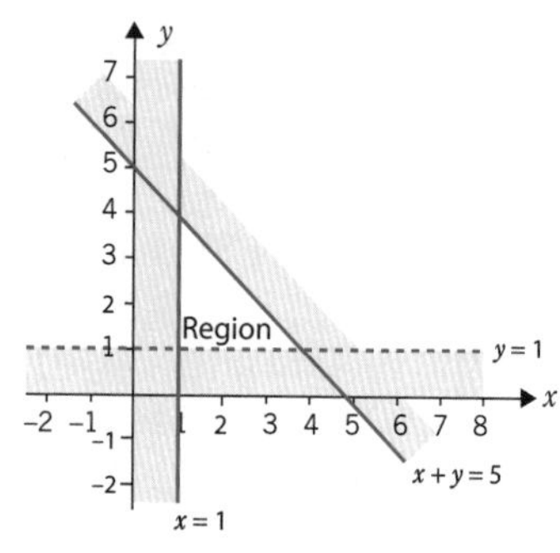

EXAM QUESTION

1. a. −2, −1, 0, 1, 2, 3, 4 b. $x < 2$

pages 46–47

Straight-line graphs

PROGRESS CHECK

1. a.

x	−2	−1	0	1	2	3
y	−1	1	3	5	7	9

b.

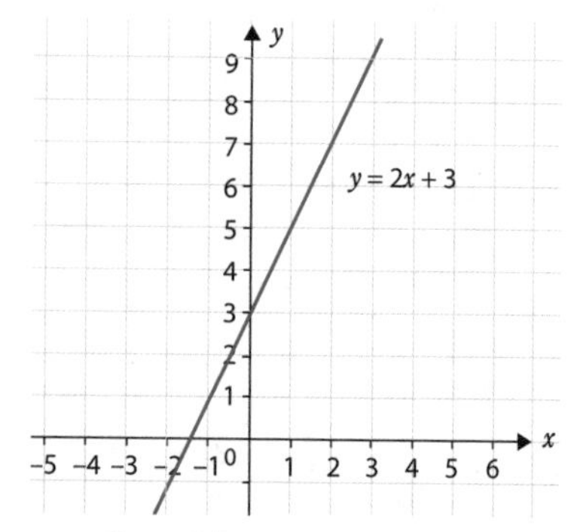

2. $y = -2x + 10$

EXAM QUESTION

1. a. $y = -2x + 3$ b. $y = \frac{1}{2}x + 3$

pages 48–49

Curved graphs

PROGRESS CHECK

1. a.

x	−3	−2	−1	0	1	2	3
y	−28	−9	−2	−1	0	7	26

b.

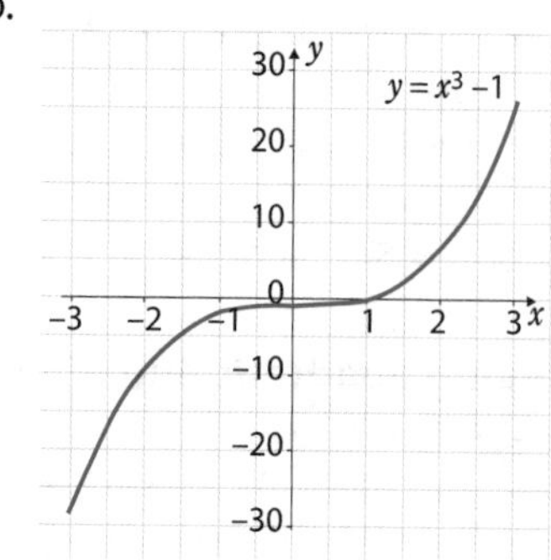

c. $x = 2.5$

EXAM QUESTION

1. Graph A: $y = \frac{3}{x}$

   Graph B: $y = 4x + 2$

   Graph C: $y = x^3 - 5$

   Graph D: $y = 2 - x^2$

pages 50–51

Interpreting graphs

PROGRESS CHECK

1. True for both solutions
2. False for both solutions
3. True for $x = -0.7$, false for $x = 5.6$

EXAM QUESTION

1. $a = 9000$ $b = \frac{2}{3}$

pages 52–53

Functions and Transformations

PROGRESS CHECK

1. a

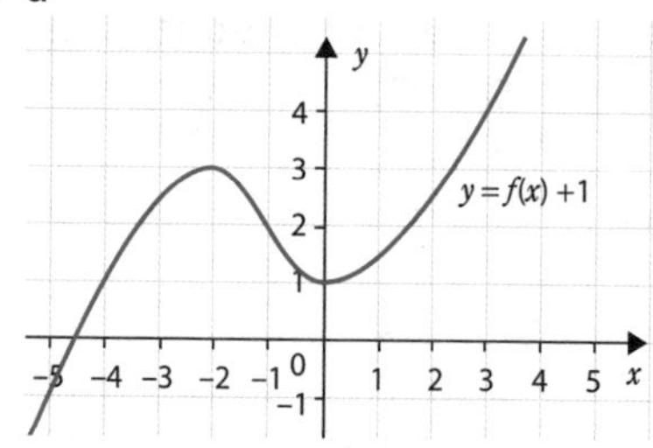

b

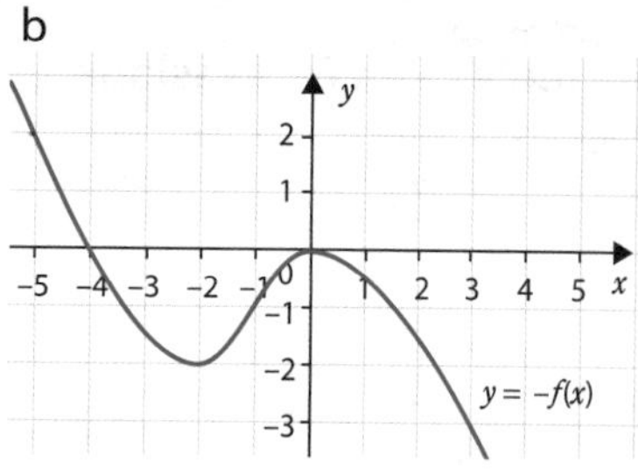

EXAM QUESTION

1. (−1, −3)
2. (2, −7)
3. (4, −3)
4. (−2, −3)
5. (1, −3)

Day 5

pages 54–55

Constructions

PROGRESS CHECK

1. Construct an equilateral triangle first. Bisect the 60° angle.

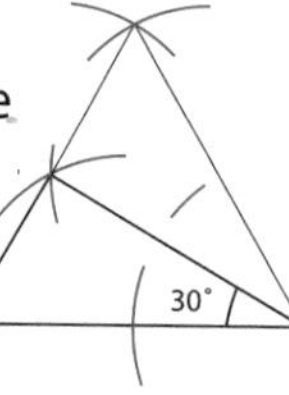

2. Perpendicular bisector of an 8 cm line drawn.
3. Angle bisected with compasses only.

EXAM QUESTION

1.

a.

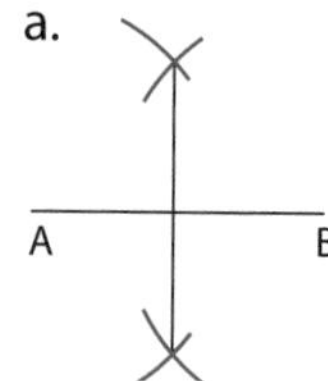

b.

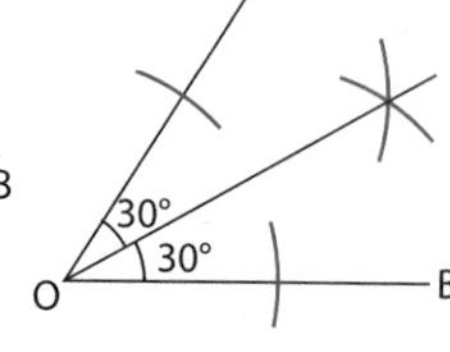

pages 56–57

Loci

PROGRESS CHECK

1.

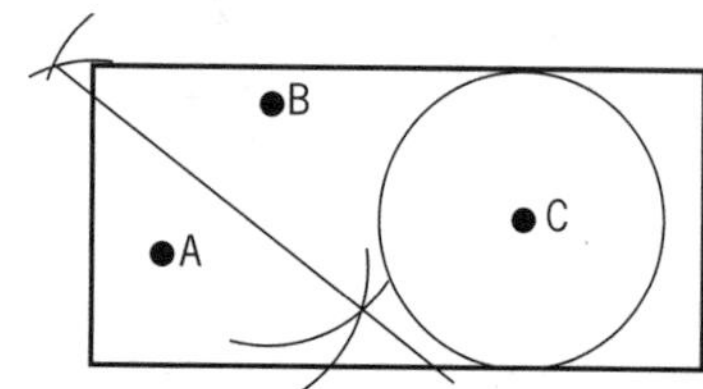

EXAM QUESTION

1.

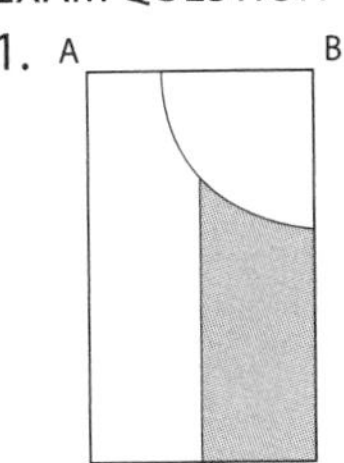

pages 58–59

Translations and Reflections

PROGRESS CHECK

1. Reflection in $y = 0$ (x-axis)
2. Reflection in $x = 0$ (y-axis)
3. Translation of $\begin{pmatrix} -6 \\ 0 \end{pmatrix}$
4. Translation of $\begin{pmatrix} 5 \\ -1 \end{pmatrix}$

EXAM QUESTION

1. a. b. see below

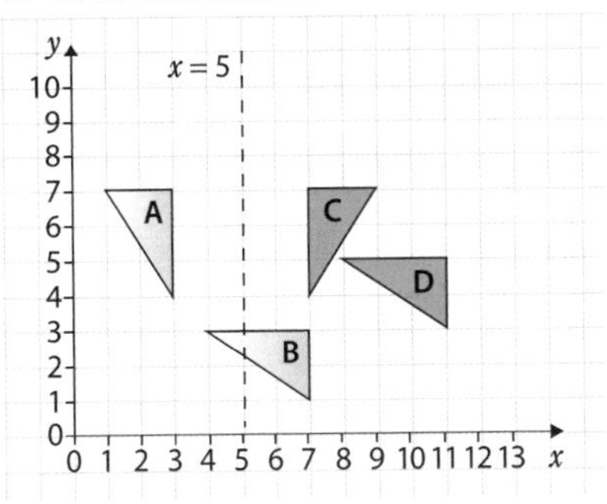

c. Reflection in the line $y = x$

pages 60–61

Rotation and Enlargement

1.

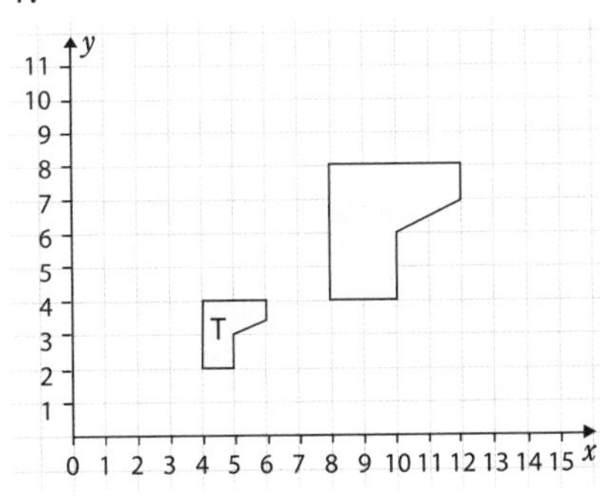

EXAM QUESTION

1.

pages 62–63

Similarity

PROGRESS CHECK

1. a. 20.9 cm b. 13.8 cm
   c. 4.5 cm

EXAM QUESTION

1. 28.4 cm^2 (3sf)

pages 64–65

Circle theorems

PROGRESS CHECK

1. a. 62° b. 109° c. 53° d. 50°
   e. 126°

EXAM QUESTION

1. a. 19° A tangent and radius meet at 90°
   b. 71° Alternate Segment Theorem or angle in a semicircle is 90°
   $E\hat{F}G = 90°$. Hence $G\hat{E}F = 71°$.

pages 66–67

Pythagoras' Theorem

PROGRESS CHECK

1. Since $26^2 = 24^2 + 10^2$
   $676 = 576 + 100$ and this obeys Pythagoras' theorem, then the triangle must be right-angled.

EXAM QUESTION

1. 15.3 cm
2. $\sqrt{149}$

Day 6

pages 68–69

Trigonometry

PROGRESS CHECK

1. a. 5.79cm b. 4.64cm
   c. 8.40cm d. 24.11cm
2. a. 38.7° b. 40.4° c. 43.0°
   d. 24.8°

EXAM QUESTION

1. a. 57.4° (3sf)
   b. $\text{Sin } 48° = \frac{19.7}{PA}$ $PA = \frac{19.7}{\text{Sin } 48°}$
   PA = 26.5 m (3 sf)

pages 70–71

The Sine and Cosine rules

PROGRESS CHECK

1. a. Correct b. Correct
   c. Incorrect
2. a. 114° (nearest degree)
   b. 53° (nearest degree)

EXAM QUESTION

1. 16.4 m (3 sf)

pages 72–73

Trigonometric functions

PROGRESS CHECK

1.

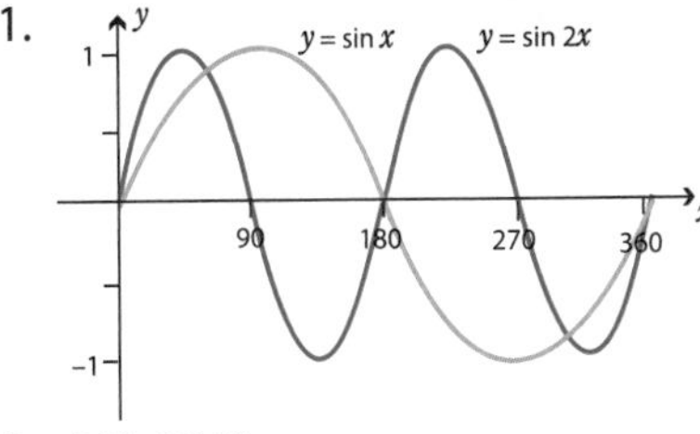

2. 60°, 300°

EXAM QUESTION

1. –225°, 45°, 135°
2. 30°, 330°

pages 74–75

Arc, sector and segment

PROGRESS CHECK

1. a. 14.05 cm b. 49.17 cm^2
   c. 26.97 cm^2
2. True since
   $\left(\frac{70°}{360°} \times \pi \times 8^2 - \frac{1}{2} \times 8 \times 8 \times \sin 70°\right)$
   $= 9.03$ cm^2

EXAM QUESTION

1. 30 cm^2

pages 76–77

Surface area and volume

PROGRESS CHECK

1. 565 cm^3 2. 637 cm^3
3. 905 cm^3 4. 245 cm^3

EXAM QUESTION

1.

$\frac{1}{3}bd$	πc	$2d$	bcd	$c(b-d)$	$\frac{cd}{b}$	$\frac{\pi b^2 d}{c}$
✔				✔		✔

2. Radius = 2.61 m (3 sf)

Day 7

pages 78–79

Vectors 1

PROGRESS CHECK

1.

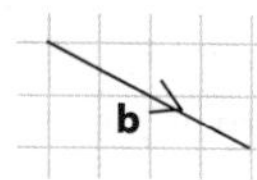

2. Yes, vectors **c** and **d** are parallel since **d** = 2**c**
3. a. $\sqrt{4^2 + 2^2} = \sqrt{20}$
   b. $\sqrt{(-3^2) + 4^2} = \sqrt{25} = 5$
   c. $\sqrt{(-2)^2 + 0^2} = \sqrt{4} = 2$
   d. $\sqrt{6^2 + (-1)^7} = \sqrt{37}$

EXAM QUESTION

1. a. $\begin{pmatrix} 4 \\ 5 \end{pmatrix}$ b. (5,4)

pages 80–81

Vectors 2

PROGRESS CHECK

1. a. i. $-\mathbf{a} + \mathbf{c}$ ii. $-\mathbf{c} - \mathbf{a}$
   b. $\mathbf{a} + \frac{1}{2}(-\mathbf{a} + \mathbf{c}) = \frac{1}{2}\mathbf{a} + \frac{1}{2}\mathbf{c}$
   or $\frac{1}{2}(\mathbf{a} + \mathbf{c})$

EXAM QUESTION

1. a. $-\mathbf{a} + \mathbf{b}$ b. $-2\mathbf{a} + 2\mathbf{b}$
2. $-\mathbf{a} + \mathbf{b}$
3. Since $\overrightarrow{FE} = -\mathbf{a} + \mathbf{b}$ and $\overrightarrow{AB} = 2(-\mathbf{a} + \mathbf{b}) = 2\overrightarrow{FE}$
   Vector $\overrightarrow{AD}$ is twice the length and parallel to $\overrightarrow{FE}$.

pages 82–83

Scatter diagrams and correlation

PROGRESS CHECK

1. true 2. false 3. true
4. true 5. true

EXAM QUESTION

1. a. Correctly drawn line of best fit
   b. approx. 1 year old
   c. approx. £3 000

pages 84–85

Averages

PROGRESS CHECK

1. 150.9 2. 2, 2.7, 4, 5

EXAM QUESTION

1. 5.52 hours

pages 86–87

Cumulative frequency graphs

PROGRESS CHECK

1. i. 9B
   ii. Class 9A. The median is higher than 9B, 50% of students in 9A scored over 64 marks. The top 50% of students in class 9B scored over 55 marks.

EXAM QUESTION

1. a. i. lower quartile = 12
   ii. interquartile range = 11
   b. .

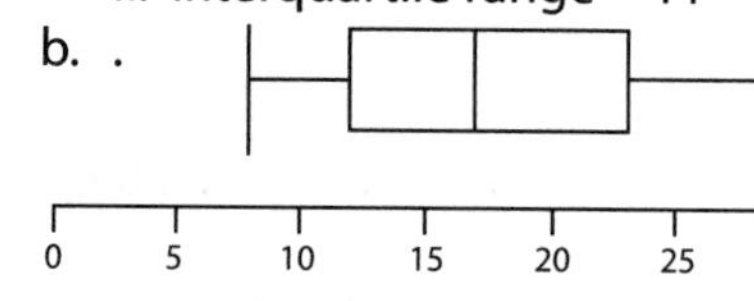

pages 88–89

Histograms

PROGRESS CHECK

1.

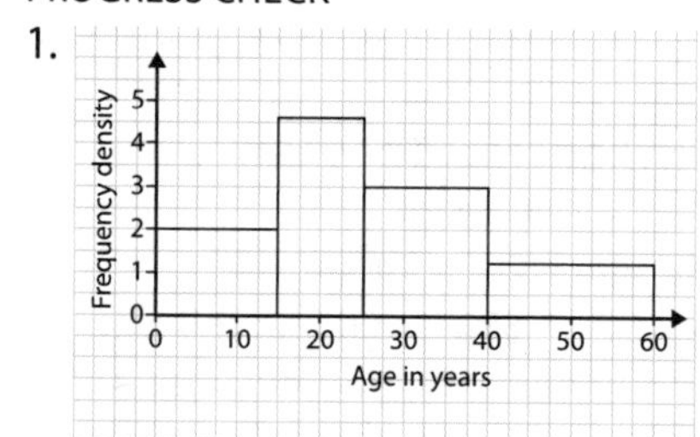

EXAM QUESTION

1.

Distance (km)	Frequency	Frequency density
$0 < d \leqslant 15$	12	0.8
$15 < d \leqslant 25$	32	3.2
$25 < d \leqslant 30$	36	7.2
$30 < d \leqslant 45$	60	4.0
$45 < d \leqslant 55$	20	2.0

2.

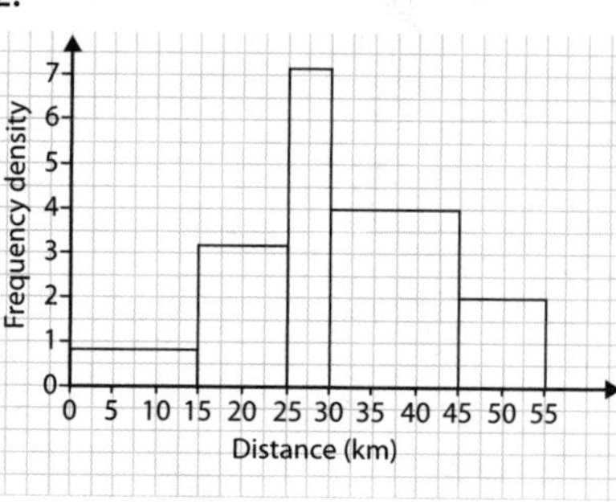

pages 90–91

Probability

PROGRESS CHECK

1.

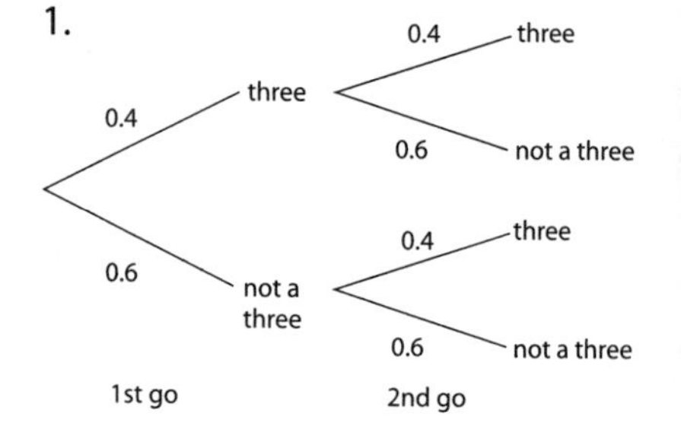

2. a. 0.16 b. 0.48

EXAM QUESTION

1.

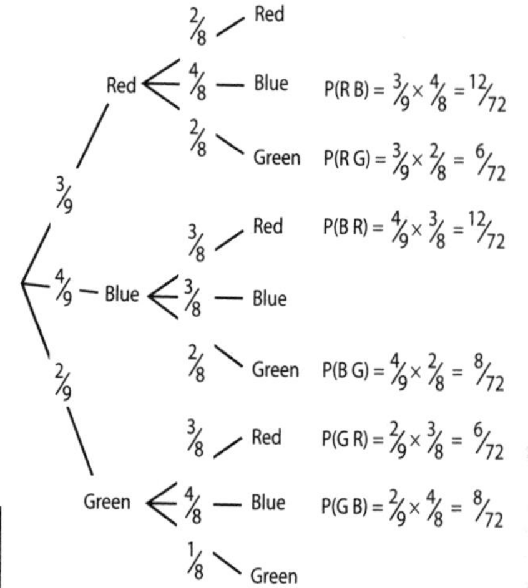

P(two different colours) = $\frac{52}{72}$

Or 1 – P(same colours)

$= 1 - \left(\left(\frac{3}{9} \times \frac{2}{8}\right) + \left(\frac{4}{9} \times \frac{3}{8}\right) + \left(\frac{2}{9} \times \frac{1}{8}\right)\right)$

$= 1 - \frac{20}{72} = \frac{52}{72}$